SPORT, LEISURE AND CULTURE IN THE POSTMODERN CITY

Heritage, Culture and Identity

Series Editor: Brian Graham,
School of Environmental Sciences, University of Ulster, UK

Other titles in this series

Sport, Leisure and Culture in the Postmodern City

Edited by

PETER BRAMHAM and STEPHEN WAGG
Leeds Metropolitan University, UK

ASHGATE

Published by
Ashgate Publishing Limited
Wey Court East
Union Road
Farnham
Surrey, GU9 7PT
England

Ashgate Publishing Company
Suite 420
101 Cherry Street
Burlington
VT 05401-4405
USA

www.ashgate.com

British Library Cataloguing in Publication Data
Sport, leisure and culture in the postmodern city. --
   (Heritage, culture and identity)
   1. Social change--England--Leeds. 2. City and town life--
   England--Leeds. 3. Leisure--England--Leeds. 4. Leeds
   (England)--Social conditions--21st century. 5. Leeds
   (England)--Cultural policy.
   I. Series II. Bramham, Peter. III. Wagg, Stephen.
   942.8'19086-dc22

Library of Congress Cataloging-in-Publication Data
Sport, leisure, and culture in the postmodern city / [edited] by Peter Bramham and Stephen Wagg.
       p. cm. -- (Heritage, culture and identity)
   Includes index.
   ISBN 978-0-7546-7274-6 (hardback) -- ISBN 978-0-7546-9102-0
(ebook) 1. Sports--Social aspects--England--Leeds. 2. Leisure--Social aspects--
England--Leeds. 3. Industrialization--England--Leeds. 4. Urbanization--England--Leeds.
5. City planning--England--Leeds. I. Bramham, Peter. II. Wagg, Stephen.
   GV706.5.S7345 2009
   306.4'830942819--dc22

                                                                                        2009030836

ISBN: 9780754672746 (hbk)
ISBN: 9780754691020 (ebk)

Mixed Sources
Product group from well-managed
forests and other controlled sources
www.fsc.org Cert no. SGS-COC-2482
© 1996 Forest Stewardship Council

Printed and bound in Great Britain by
TJ International Ltd, Padstow, Cornwall

Contents

List of Contributors

Each contributor teaches, or has taught, at Leeds Metropolitan University – all bar two in the Sport and Leisure Teaching and Research Group in the Carnegie Faculty of Sport and Education.

Peter Bramham is currently Visiting Research Fellow at Carnegie Faculty of Sport and Education and author of a number of books on leisure.

Ben Carrington completed his doctorate at Leeds Metropolitan University, where he is a Carnegie Visiting Research Fellow. He teaches sociology at the University of Texas at Austin.

Janet Douglas is a Principal Lecturer in Politics in the School of Cultural Studies.

Jonathan Long is a Professor in the Carnegie Faculty.

Aarti Ratna is a Lecturer in the Carnegie Faculty.

John Spink retired from a Senior Lectureship in Carnegie in 2005.

Karl Spracklen is a Principle Lecturer in the Carnegie Faculty.

Ian Strange is a Professor and Director of the Centre for Urban Development and Environmental Management.

Stephen Wagg is Professor of Sport and Society in the Carnegie Faculty.

Chapter 1

Introduction

Peter Bramham and Stephen Wagg

> There is a strong predilection these days to regard the future of urbanization as
> already determined by the power of globalization and of market competition.
> Urban possibilities are limited to mere competitive jockeying of individual cities
> for position within the global urban system. Harvey, D. (1996: 420)

Both in his seminal book *Conditions of Postmodernity* (1989) and in an earlier
article, 'Down towns' published in *Marxism Today* (1989) David Harvey argued
that late capitalism results in the 'serial reproduction' of malls, pedestrian city
centres, plazas and waterfronts as capital markets reinvest and restructure,
moving away from industrial production and shifting into the service sectors,
financial services, marketing and retailing. Global firms in retailing, tourism, hotel
accommodation and fast-food outlets have invested in urban prime sites with
the result that the mix of shopping and leisure experiences varies little from one
city to another. This has led some commentators such as George Ritzer (2004) to
suggest particular distinctive places disappear only to be replaced by universal
homogenous 'non places'. Everywhere is nowhere; all places are pretty much the
same. Globalisation gradually erases distinctive localities and local identities.

This book provides an explicit focus on sport, leisure and culture because local
politicians and policy communities in recent years have identified these areas as
crucial sites for public policy and local agency. In the UK the local state has been
'hollowed out' by the centralising policies of Thatcherite and Blairite regimes.
So in the face of globalisation and centralisation, cities look towards postmodern
cultural forms, branding and lifestyles to differentiate themselves from other
cities.

In the UK experience, London Docklands and Canary Wharf represent the
archetypal postmodern city: conceived by a central government quango with
planning powers and budgets to reconstruct and redevelop derelict industrial
locations into commercial waterfronts. The vision is one of private enterprise,
high-tech, and an integration of work, home and leisure within the city. Its icons
are postmodern architecture, a mass transit system, water-based developments
and heritage conservation (retaining dockland cranes for aesthetic purposes, for
example), alongside cultural investments in cinema, art galleries and festival
events.

The backdrop to such urban redevelopment remains in social divisions of class,
race, gender and locality. What is to be done with existing local working-class

communities rooted in conditions of industrial modernity? Where are they in terms of employment, housing and leisure? How do local established industrial lifestyles shaped by race and gender fit the new cosmopolitan postmodern city? What are the politics of the postmodern city and what public policy do local politicians favour? One of the major problems associated with urban redevelopment is that local people and communities demand involvement, democratic participation and may actively resist these postmodern plans and cosmopolitan 'new times'. In the UK, Urban Development Corporations, harbingers of a neo-liberal project to energise market forces in the 1980s, were for the most part unprepared for the resistance of people in London's Docklands and have tended, since Docklands, to favour developing 'people-less' areas elsewhere.

London Docklands highlighted the pressing dilemmas of urban development with all its usual suspects, both on centre stage and lurking in the wings. Tensions in the postmodern city have played out in the face of a changing economic climate, whether the credit booms of the 1980s, the fears of recession in the 1990s or the spectre in the 2000s of full-blown global depression. Major players tend to act out their stereotypical roles: footloose international capital searching out profitable locations and niche markets; central government providing subsidies for relocation[1]; and local involvement stage-managed by local politicians in *ad hoc* partnerships and pragmatic alliances, demanded by new business models.

But there is also a growing voice of 'white racism', with the British National Party winning local elections on Isle of Dogs and gaining footholds elsewhere. Here blossom fascist parties which address fears and concerns of local working-class communities which 'dare to say what you are thinking',[2] seeking "to withdraw from the EU, protect British jobs for British workers and to say no to immigration". Even before the May 2009 media and public outrage at MP's expenses and accusations of sleaze, mainstream parties had been unable to mobilise an increasingly disenchanted electorate, particularly at a local or more global European level. During the past decades in its quest to modernise, the Labour Party has redefined its relations with traditional working-class communities; it has reneged on Old Labour's shibboleths in defence of welfarism and social reformism. One consequence of the neo-liberal project and its global politics of marketisation and privatisation has been emergence of single issue, community-based groups to promote or resist new developments, many formed in response to cuts in local public expenditure on collective consumption, often informed by an environmentalist perspective to tackle broader 'green' issues related to overpopulation, global warming and overconsumption.

---

1   See for example the Department of the Environment guaranteeing occupancy of Canary Wharf and regional authorities such as Yorkshire Forward providing funding to build the Leeds Arena in 2012 much to the dismay of politicians in competitor and neighbouring cities such as Sheffield.

2   See for example the headline slogan of the BNP in the June 2009 European Elections for the Yorkshire and Humberside region.

Manuel Castells' (1977) (1978) initial neo-Marxist analyses were overoptimistic about the emergence of new urban social movements which he predicted would shift agendas and create a 'new politics' of identity, reflecting issues about gender, race and 'green' lifestyles. Not only did those in receipt of public subsidy (in essence the powerless, marginalised and unemployed) fail to mount a defence of collective consumption expenditures in the face of retrenchment from the neo-liberal project but neither did local government professionals and public sector unions. Some cities did resist central government 'interference' and cuts in local state autonomy whereas others did not and were able to accommodate local policies within the thrust and direction of neo-liberal or New Right central government policies. Indeed Ian Henry (2001) has written extensively about Right- and Left-Post Fordist policies to deal with these times and the tensions between the central and the local state. Whereas traditional Fordist policies defended high levels of expenditure on collective consumption, Right-Post Fordism accepts the central state's neo-liberal project and looks to the local state to be more entrepreneurial and managerial, contracting services out to the most efficient and effective provider. In contrast Left-Post Fordism, demands that the local state offers progressive new cultural services for a 'rainbow alliance' of excluded and neglected citizens, such as racial and ethnic minorities, gay and women's groups.

It is clear from Peter Bramham and John Spink's first chapter here about postmodern cities and the detailed city-centre case study provided in the second chapter by Janet Douglas, that Leeds is just one of many northern cities that have experienced the Docklands strategy: namely, the redevelopment of waterfronts, derelict warehouses, canal sites into an integrated city-centre environment. There are other examples in northern UK cities with Manchester, Liverpool, Sheffield and smaller industrial towns Bradford, Wakefield, Halifax, Barnsley and Doncaster in South and West Yorkshire.[3]

Leeds became a postmodern city in its patterns of employment and labour migration as the past generation for thirty years witnessed a decline in primary and secondary sector jobs and a growth in tertiary sector employment in education, insurance and financial services, tourism, arts and cultural services. All these were designated as 'qualities' in the City's new unitary plan *A Vision of Leeds*, first published in 1992.[4] Hence, there were several zones or prime use quarters: Civic Quarter, Prime Office Quarter, Prime Shopping Quarter, Hospital Quarter, Education Quarter, Riverside Area and Prestige Development Area. There were key developments and the areas of city which symbolised these dimensions of

---

3 West Yorkshire's five towns are Leeds, Bradford, Huddersfield, Wakefield, and Halifax – all at the heart of industrial modernity with local economies dominated by coalmining, textiles, engineering, printing and railways.

4 This Vision has been renegotiated with the then new leader of the Leeds City Council, Brian Walker, putting out the original plan to be renegotiated and relegitimated by a new consultative process in 1997 and since then has updated – see for example, Leeds Unitary Development Plan (June Review 2006) and the Vision for Leeds 2004-2020.

economy and policy – such as Lisbon Court; Corn Exchange; St. James Hospital and Thackray Museum; the University; Canalside; Armouries; West Yorkshire Playhouse area and so on.

Since 1979 the city administration had until 2004 been controlled by a Labour council and the local economy had remained buoyant compared with other northern industrial towns. Nevertheless, a Thatcherite central government imposed an Urban Development Corporation on the city during 1987-1995 with an annual £15 million budget and planning powers to encourage capital investment and regeneration in inner-city areas (free from some local authority professional and bureaucratic planning control). The slogans were for 'new realism', entrepreneurialism, leaving the private sector to develop those areas which had been 'failed' by local authority red tape, blanket subsidies and inadequate market intelligence and data. The Labour Group responded to the UDC in Leeds by setting up its own Development Corporation (Leeds City Development Corporation) and transferring its land ownership and assets so as to isolate or minimise the impact of the Leeds UDC. The Board of the LCDC included a partnership of Chamber of Commerce, local politicians and leading experts and planners.

The traditional service committee structure and divisions of local authority officers and departments experienced substantial changes. Like other local authorities, Leeds was subject to new legislation from central government (e.g. Rate Capping, Community Charge, Urban Development Corporation, Compulsory Competitive Tendering) to curtail or restrict local powers of expenditure and policy direction. Leeds' local political system, whether controlled by the Conservatives or Labour, has historically been dominated by a 'rate-payer' ideology. Here, emphasis is focused on constraining budgets, gaining accountability and value for money, and political pragmatism in response to local political demands and pressure groups. The Leeds policy system was already well-positioned to develop Right-Post Fordist strategies to go with the political plans of central government policies.

Leeds was already seeking strong partnerships with the commercial/financial sectors of the local economy. It had set up Leeds Waterfront in the early 1990s with a £500K budget to develop the waterfront and increase tourism spending by 12% and visitor numbers by 10% in the three year period. The Leeds Initiative was another attempt to co-ordinate the public, private and voluntary sector response to redevelopments, events and festivals. It was headed by a flexible taskforce (and initially using Urban Aid money from central government), it introduced 'Landmark Leeds', which sought to conserve heritage architecture in Leeds and refurbish or reconstruct central city Victorian Leeds street furniture. It was also involved in urban event 'pump priming' such as sponsorship for City of Flowers, for festivals – film, jazz, piano as well as co-ordinating events like Opera North, rock concerts in the park – in the early days boasting global celebrities such as Madonna, Michael Jackson and the Rolling Stones, but recently with more local celebrities, such as Arctic Monkeys and Kaiser Chiefs topping the bill at the Leeds Festival.

There has been discussion amongst urban theorists regarding city tourism and 'boosterism' in postmodern debates. David Harvey has argued that the relocation and restructuring of city forms around the post-modern is a 'carnival mask' to conceal deep-rooted class divisions with ensuing privatisation of social, cultural and political forms and spaces. If post-modern buildings are the precursor of postmodernity, then Leeds has its fair share. Indeed, architects refer to the 'Leeds style' – decorative brick, a bricolage of styles Greek, Gothic, Victorian – glass, steel and brick. Traditional buildings from industrial modernity have been transformed into postmodern restaurants and conference centres. The Corn Exchange has been refurbished as exclusive niche market shops, Granary Wharfe becomes 'specialty shopping' – 22 highly individual gift shops, art and craft stalls in a festival market and new cultural forms have emerged such as the West Yorkshire Playhouse, Opera North and the Phoenix Dance Theatre boasting an international reputation for audience development and community dance education.

Leeds offers a Chicago skyline, but the facades of modern industrial functional office space have been reclad in post-modern glass and colours (see for example Leeds Metropolitan University Library and the iconic Rosebowl which houses the Leeds Business School and opened in May 2009), shopping mall and precincts have emerged such as Bond Street Centre and the Victoria Quarter and the city council has taken ownership of the Elland Road complex, home of Leeds United's football ground which has been refurbished to house national and European competitions.

The success and appearance of post-modern buildings have helped convince city business and politicians that revitalisation and urban renaissance are on the way and civic boosterism is actually working. City politicians take credit for developments that would have happened anyway, and claim that changing cityscapes with new investments, especially new leisure opportunities, benefit every citizen. The postmodern discourse persuades doubters that it attracts new businesses. Newcomers to the region and locals themselves, it is argued, benefit from new employment, new buildings and a refurbished environment. Within these policy communities, urban marketeers convince themselves and others that they have made a 'difference'. Indeed, the key Leeds Urban Development Corporation slogan was '*Making Things Happen*' – *The Delivered Future*, but in reality, many UDCs have failed (and all have been wound up) and the Leeds UDC may have taken credit for planning processes which were embryonic and agreed *before* their incorporation. City politicians and policy communities have been swept along with the euphoria of urban renaissance, pressing for urban involvement and New Right-Post Fordist developments, even though they have little concrete evidence that festivals, heritage events, sports and cultural industries and such like generate extra investment and are central factors in relocations. Politicians were prepared to believe these were important issues in the city's strategy and in consultative documents. City boosterism and mega-events become important ingredients in city politics, leisure and tourism strategies. This policy shift is perhaps synonymous with a significant loss of civic power and municipal effectiveness in more important areas of urban life. Maurice Roche (2000) has argued that the benefits of such sports and leisure strategies have been problematic and his case study (Roche 1994) of the Sheffield Student Games in the early 1990s convincingly argued that both conventional and situational rationality in local political processes have been the first casualties when preparing, deliberating and costing civic bids.

Indeed, rather than studying tourist multiplier effects of mega-events, there should be more detailed empirical research on political configurations that lead to such decision-making. The third chapter in this book by Jonathan Long and Ian Strange makes such a contribution to understanding the development of cultural policy in Leeds. It has often been that it was political 'autarchies' and struggles for power within city policy systems that best explained decision-making processes and outcomes, rather than other political, economic or cultural assessments.

The dominant Leeds Labour group has not been immune to post-modern visions of using leisure and cultural policy as part of urban regeneration strategy. In the early 1990s the Policy Resources Committee has stressed the vision of Leeds as a 24 hour European City. At that time, Council Leader Trickett mapped out his European vision to lead businesses, police and policy makers to grant 24 hour licences for bars, restaurants and discos, as well as encouraging property owners to cut rents to encourage late-night shop opening. Founded in 1990 the Leeds Initiative encouraged local partnerships to revitalise Leeds city centre, giving grants for illuminating buildings, street events, festivals, Valentine Fairs in the city centre, as well as Christmas Lights, City Centre Cycling, Kellogg's Tour of Britain and Leeds Classic Cycle Race. This bricolage of leisure and cultural activities is not part of the generic sports, parks, culture and leisure policy but has been organised on a more flexible, pragmatic, annual basis through contingency funds from the Leader's Office. There has been a pragmatic alliance of existing provision and new projects initiated by civic personalities and professional officers. There have been new opportunities in these new times which are heralded at a local and national level, with key policy makers looking towards European cities and cultural renaissance, whilst professional officers remain still quite local and provincial. In the past there has been a local policy officer system with an emphasis on the 'Leeds' way of doing things within a stable and safe 'rate-payer' ideology. So arose an unresolved tension between traditional parochial Leeds style 'mass' leisure service provision associated with previous Leeds Leisure Services and emergent demands from the centre for a vibrant 24hour tourist city. There remains what Raymond Williams called 'a local structure of feeling'.[5] It is this culture that has been celebrated in the writings of Alan Bennett, one of Leeds' few national celebrities. His postmodern celebrity status is discussed by Peter Bramham in the fourth chapter of this collection. The old Leeds industrial culture of modernity

---

5   This concept was deemed central to understanding Manchester and Sheffield see Taylor, L., K. Evans, et al. (1996). *A tale of two cities: global change, local feeling and everyday life in the North of England – a study in Manchester and Sheffield* London, Routledge.

was going home for 'tea' (the main meal of the day) after work rather than more flexible work patterns around the city-centre restaurants, pubs, bars and cafes in the new 24/7 postmodern city. A similar cultural clash took place when Leeds United entered UEFA Cup competitions for the first time in 1966; programme notes explained to Leeds United supporters precisely how to pronounce strange sounding football teams and where in Europe they were located.[6] It is precisely this juxtaposition of the local and the global, the modern and the postmodern that is explored both in Stephen Wagg's chapter about the Revie era of Leeds football and Karl Spracklen's chapter about rugby league. Local supporters belong to and remain loyal to local teams such as Hunslet and Bramley despite the success of Leeds Rhinos sustained by the media exposure, branding and marketing of Super League.

It is therefore essential for any analysis of the evolving city to present an explanation with a precise level of detail and the present chapters focus on the diverse contributions of sport, leisure and culture. Broad structural processes have been identified in various literatures about economic restructuring, the relocation of spatial forms, cultural change and social divisions. If it is the case that economic, political, social and cultural formations have their distinctive trajectories then these dimensions need to be traced spatially and historically within city case studies. The demands of city images, new experiences, heritage and style focus attention on the construction and rediscovery of new prime sites. These represent the post-modern city and provide raw materials for constructing new narratives about changing spaces and places. It is important to examine historically the process whereby new locations and new forms are constructed within the context of changing economic, political, cultural and social formations. Standing in Millennium Square, Leeds Shopping Plaza, Leeds Waterfront, Temple Newsam or in Roundhay Park, one not only needs to exercise the tourist gaze on new installations, the multiplex cinema, the themed garden and visitor centre, the museum shop and art gallery exhibitions, but also to ask questions about who is exercising the gaze and whose interests are served by such developments.

Such questions come to light in Ben Carrington's chapter which draws upon personal recollections and reflections of key moments of coming to and becoming an outsider in the city, experiencing the dark side of Leeds' 'local structure of feeling' whether through the eyes of the child in Kirkstall, playing cricket in Chapeltown or working at Leeds Metropolitan University. The penultimate chapter by Aarti Ratna's explores issues of negotiating black and Muslim identity in young women's sporting lives, particularly in relation to playing representative football. The postmodern may celebrate difference but its contours are projected onto existing unequal power relations of race and ethnicity. It may feel like postmodern concepts of bricolage, dedifferentiation, hybridity and difference has long been part of social science discourses about the city but there has still been little by way of

---

6   In the 1966 season, Leeds played such teams as Torino, Leipzig, Újpest Dózsa, Valencia and Real Zaragoza.

critical questioning about the trajectory of particular cities. The following chapters on Leeds will, it is hoped, make some contribution to developing a nuanced and critical analysis of the postmodern city.

References

Castells, M. (1977). *The Urban Question*. Cambridge, Mass. and London, MIT Press, Edward Arnold.

Castells, M. (1978). *City, Class and Power.* London, Macmillan.

Harvey, D. (1989). *The Condition of Postmodernity: an Enquiry into the Conditions of Cultural Change*. Oxford, Blackwell.

Harvey, D. (1989). Down Towns. *Marxism Today*, (January).

Henry, I. (2001). *The Politics of Leisure Policy.* Basingstoke, Macmillan.

Ritzer, G. (2004). *The Globalisation of Nothing*. London, Thousand Oaks, New Dehli, Pine Forge Press, imprint of Sage publications.

Roche, M. (1994). "Mega-events and urban policy." *Annals of Tourism Research* **21**(1): 1-19.

Roche, M. (2000). *Mega-events and Modernity: Olympics and Expos in the Growth of Global Culture*. London, Routledge.

Taylor, L., K. Evans, et al. (1996). *A Tale of Two Cities: Global Change, Local Feeling and Everyday Life in the North of England – a Study in Manchester and Sheffield*. London, Routledge.

Chapter 2

Leeds – Becoming the Postmodern City

Peter Bramham and John Spink

Globalisation and Postmodernity

Whether styled as the coming of Postmodernity, of Postindustrialism, of Postfordism or even of Late Capitalism, there is substantial evidence that the developed world has sustained significant and qualitative change in the past 30 years. Many authors pinpoint the mid 1970s with the onset of the Oil Crisis and ensuing global depression and fiscal crises for cities, particularly in North America, as marking a critical juncture in these developments. From that time the taken-for-granted dominance of the West came to be challenged by the emergence of newly-developing economies on the Pacific Rim. However, by the late 1990s, even the 'tiger economies' in this part of the world were experiencing substantial economic disruption, whilst transnational agencies wrestled with intractable problems surrounding world debt. During the next decade the exponential economic growth of both China and India made unprecedented demands for primary materials for energy, food production and manufacture. Recently there have been severe dislocations in both commercial and financial markets which fashioned global repercussions, signified by the 'credit crunch' following the collapse of sub-prime mortgage markets in the US and elsewhere. National governments worldwide have chosen to provide substantial public funds to help the private sector in banking, building societies, car manufacture and small business. There has also been consistent global failure to agree and meet climate change targets for the environment in 'recycling' and transport. Another major contributor to widespread destablisation was the liberalisation of state-socialist nation states with their subsequent political desire to form new military and economic links with western capitalist democracies. In the UK, media and politicians have debated the social and economic impacts of migrant labour and asylum seekers, following the expansion of the European Union from 15 to 24 members in 2004.

Most social science commentators characterise such seismic changes in ideas, communications, manufacture, information, markets and natural environment by frequent use of the umbrella term of 'globalisation'. Whatever the exact cause or its precise timing, the consequences for urban centres and for the leisure time of their inhabitants have been profound. These changes have affected all aspects of contemporary life.

Economic Change

At the core of all cities is their economic rationale. Postmodernity in the form of postindustrial relations reflects profound change in urban economies. Traditional methods of large-scale factory-based, mass production, usually organised around assembly-line production techniques, have given way to much more flexible methods. Mass production meant efficient supervision and control technologies, high levels of productivity and low unit costs of production. More recently, in place of standardised products created through economies of scale in massive plants dominating the inner areas of large industrial cities, manufacturers have adopted small-batch production methods for high-quality products using computer-aided design and manufacturing. This 'just in time' production allows for economies of scope and a far more demand-responsive system based on fewer but highly skilled workers within smaller plants that can even be located away from urban centres.

This flexibility facilitated within production systems has generally led to less routinised work and to the employment of a smaller, less unionised and increasingly part-time workforce. According to Leon Kreitzman (1999) patterns of working have become more flexible as boundaries between work and leisure become more permeable in the 24 hour society. Reduction in the old hierarchical labour force has in most developed countries been accompanied by growing pools of persistent long-term unemployment, particularly amongst less skilled male workers. Simultaneously, with change in production patterns and the application of new technologies, has come a structural shift in local and national economic life. In cities like Leeds, manufacturing industry in engineering, clothing or printing has come to assume a less significant role and in most centres has been displaced by a growing focus on commerce, services and finance capital. Growth in this service sector has been facilitated by technological developments in 'global information highways' viz. the digitalization of computing, data-processing and satellite and cable communications. According to Manuel Castells (1996) we live in a 'network society' at 'an interval in history' which is experiencing a transformation of material culture by new technologies of information processes and communications. This 'lifeblood' of cities demands locations near central hubs or nodes of information flows in order to avoid the risk of being bypassed or 'switched off', as is the plight of undeveloped regions at both a national and international level. Hotspots of commerce, finance and information transfer, whether in West Coast US, in Northern Italy, or the South East of England, must be attractive to dominant managerial elites who express cosmopolitan tastes in work, residence and leisure.

The major urban changes of the past 30 years, however categorised, be it as deindustrialisation, post-modernity, high modernity or late capitalism, involved considerable dislocation and readjustment within European cities, in their economic, cultural, spatial and political structures. The generic changes which have transformed the working-class cities of the industrial heartlands of Europe have nevertheless been mediated by diversification in the local economy. Accordingly, the industrial and commercial face of cities in Britain and Europe, crafted by processes of production, have been reshaped and recast by newer forms of economic activity and distinctive patterns of consumption and pleasurable free time.

These represent decisive change in economic activity and in the labour force from one predominantly male and working in manufacturing to one which is increasingly female and working as the personnel of a postindustrial revolution. Such major structural changes have implications far beyond the economic domain. Most commentators see development of new information technologies as representing a distinctive stage in economic development as cities are tied into processes of 'globalisation'. There are now global 24 hour financial and information systems and therefore decisions by transnational agencies and corporations have a direct impact on cities wherever they are located in the world.

Social Change

Urban populations have necessarily been greatly affected by the broad structural re-composition of city economies. In place of a large working-class population based upon male manual labour have come service classes with very different attitudes and aspirations. The historic collective and community focus associated with each industrial locality has been replaced by a far more individualised and domestic perspective adopted by the new 'service' class.

Flexibility in employment has encouraged greater spatial mobility and a destabilising of local cultures and kinship networks. Giddens (1991) describes this process as the 'disembedding' of local cultures and patterns of life and their replacement by 'expert systems' of technical knowledge which stretch over time and space. To provide one example, local cuisine and restaurants become replaced by global chains such as Starbucks or McDonald's, and local shops and outfitters are replaced by exclusive fashion and lifestyle designers like those which constitute the Victoria Quarter in Leeds. Throughout the UK, there is growing popular and media resistance to the dominance of supermarket chains over customers, suppliers and producers. Tesco, in particular has been actively, too aggressively for many critics, acquiring and hoarding land for future stores, challenging local development plans and absorbing independent local stores in order to rebrand them as Tesco Metro or Tesco Express. Shopping malls, in particular, have become, in the words of George Ritzer (2004), 'non places' that carry no unique or distinguishing features as each mall, like the White Rose Centre, boasts the same national retail outlets and identical brands.

Households have changed simultaneously, with declining birthrates and more female employment leading to smaller families. Economic independence at all ages has increasingly been associated with independent living and growing demand for self-contained housing. There is some evidence[1] of changing gender roles

---

1 See National Statistics data from the 2001 Census at www.statistics.gov.uk/census2001/topics, accessed 28.02.2009 and Steve Coltrane's overview of 200 studies

and gender expectations in employment, marriage, family and leisure. Greater numbers of elderly people, due to lengthening life expectancy and continued independence, have generated many more single-person households. In addition, marital breakdown and new patterns of living and sexual expression have all contributed to an increasingly varied and fragmented social structure.[2] Some cities seek to use these changed identities as an important source of investment in leisure [for example, the 'Gay Village' in Manchester: Taylor et al. (1996)]. However, alongside the positive effects of restructuring, changes in traditional relationships with work have led to great areas of cities in which the casualties of these global innovations face lives of deprivation and material poverty within societies of widening disparity in wealth and socio-economic resources, as experienced in the vast estates of South and East Leeds.

Cultural Change

Global telecommunications and computerisation affecting industrial production have wrought a similarly profound effect on urban cultural forms. Satellite broadcasting, increasingly replacing terrestrially-sourced television in many homes, accelerates national and transnational acculturation in a homogenous commodification of international products and associated lifestyles. Particularly amongst the impressionable young such forms and products assume immense status capital as tide after tide of material innovation sweeps around the world, supplanting previous commercialised innovations with still more prestigious replacements. It is this global commodification and consumption which represents a qualitative shift from historic local and national cultural heterogeneity.

Amongst all age groups there is increased focus on home and domestic-based cultural forms in leisure time. The domestic hearth, whether as the venue for television, video game or more traditional activities like reading and needlecrafts, has come to replace public and collective domains of mass entertainment developed during the era of modernity. Even within households individualisation of headphones, laptops and palmtops and most recently iPhones, whether for auditory or visual pleasure, further separate leisure participants into an isolationist cellular mode. Time shifting and de-differentiation of patterns of life are facilitated by video timers, digitalized TV and the BBC iPlayer. Fast foods, 'snacking', microwaves and a variety of computer applications ensure that even meals are no longer necessarily shared group activities, whether at home or at work.

---

in Coltrane, S. (2004). "Research on household labor: Modelling and measuring social embeddedness of routine family work." *Journal of Family and Marriage* 62(4): 1208-33.

2   See for example patterns of polarisation and gentrification in Kirkstall, Leeds in Bramham, P. and J. Spink (1994). Leisure and the postmodern city. *Modernity, postmodernity and lifestyles*. I. Henry. Eastbourne, Leisure Studies Association: 83-103.

Individualisation, whether of personal mobility through private car use, of entertainment through personalised stereos and television sets, or even of education through CD Rom and the internet, has profoundly restructured the leisure lifestyles of contemporary urban populations. Through generic websites such as Bebo, Facebook, My Space and YouTube or through internet dating sites, blogs and chatrooms, communication has become more selective, customized and privatized. Many of the old collective forms (Hoggart 1957) enjoyed by a mass public have been eliminated or replaced by newly repackaged commodification, whether as multi-screen cinemas or executive-boxed football stadia or downloaded music and television programmes.

The postmodern emphasis placed on pleasure, hedonistic experience and consumption has across Europe revolutionised leisure and its defining forms so that shopping, travel, even physical exercise, all fit within postmodern conceptions of commodified leisure lifestyles. The focus on commercialisation and high quality has necessarily made many of these highly individualised forms elitist, exclusive and divisive, incorporating only some of the mass of urban populations. Such widening disparities in leisure opportunities remain a growing challenge for postmodern cities and in particular for urban leisure policy makers.

As one would expect, there have been substantial formative and reactive shifts in cultural policy and planning – from the 1960s counter cultural movements driven by feminism, ethnic minorities and gay liberation – all offering a redefinition of citizenship, political participation and community networks. Environmentally, there has been growth in community planning, neighbourhood centres, the recasting of public space, both in terms of safe defensible spaces and public transport, partly afforded by re-introduction of trams and mass transit systems such as in Nottingham, Newcastle, Manchester and Sheffield. There was some attempt at a new politics of representation, often through subversive cultural forms and formats in video, rock music, popular art, festivals and street events, all aimed at 'reclaiming the city' and 'Making Space'. The past twenty years have witnessed for some city planners the emergence of the 'compact city', with a pressing need for local city marketing, for improving public transport, developing brown field sites, and resisting the new urban sprawl of eco-towns and suburbs . In architectural terms this has often meant the rediscovery of the piazza, as is the case in Leeds with the opening of Millennium Square with the aid of National Lottery funding.

A good example of cultural regeneration can be seen in the city of Sheffield in the mid 1980s. After the collapse of a Labourist strategy of direct subsidy and provision for heavy industry, Sheffield sought to develop cultural industries in a 'cultural quarter' which had an arts centre, exhibition spaces, restaurants as well as recording studio for pop music, led by local bands such as the Human League and Def Leppard. The recording studio, Red Tape, was also a venue for local groups to rehearse and record, as well as operating as a training centre for unemployed youth. This theme was sustained as the city later developed the ill-fated National Museum of Popular Music.

Other cities have also developed cultural strategies. Bilbao has been faced with the rapid deindustrialisation of its primary infrastructure and socialist-led city politics within the Basque region seeks to regenerate the city as a leisure and tourist location. There have been substantial efforts to deal with river pollution and to develop and rebuild waterfront locations and invest in a modern transport system. Cultural policies have sought to attract substantial international investment viz. the Guggenheim Museum of Modern Art, refurbishment of the opera house, as well as the revitalising of local Sunday markets, festivals, fallas etc. The role of the university and education sector has been of importance, especially in attracting international conferences and reasserting Basque regional identity in Spain and in EU.

In Rotterdam, the city has re-imaged itself from a dull industrial city to a vibrant 'Water City' with diverse cultural and sporting investments – musea, 'Tropicana' sports and leisure facilities, Center Parcs, IMAX cinema, as well as substantially modernised Central Business District investment in architecturally significant office and retail buildings, with pubs, restaurants and World Championship events used to promote the 'new' city. Similarly, in Manchester, city politicians have been keen to redevelop historic buildings (the G-MEX, waterfront docklands and waterways) as well as to capitalise on club life and pop music, graphic arts scene, linked to the education sectors. Club life and the 'Gay Village' have been important destinations for new commercial investments. There has also been substantial public investment in a mass transport system (an integrated light railway in the city centre), and there is some evidence that commuters have shifted from private car usage to public transport. Manchester has celebrated its distinctive cultural forms (Manchester United and Granada Television studios) as well as mounting successive bids to hold Commonwealth Games and the Olympics.

Political Change

Growing individualism in the workplace and in home life inevitably found expression in the political realm. The old class-based workplace politics of the era of modernity reflected an expectation of a steady living in life-long production-based industries favouring unionised solidarity and a collectivist approach to political and social issues. This produced in the period after 1945 the collectivist welfare interventions of large and powerful bureaucratic states aiming at economic growth and assured living standards for mass urban populations. The precise settlement agreed between state, private capital and the labour movement was historically negotiated and contested within each nation state in Europe. The overall settlement was generically the development of postwar welfare states (or what some writers term Fordist systems of regulation), with US military, political and economic support.

During the 1980s and 1990s politics in Europe moved towards monetarist New Right policies to control public and welfare expenditure; this was known as Thatcherism in the UK, but many other nation states, France, Germany,

Netherlands, Belgium, and Spain took a similar course. During the late 1990s the European Union has sought political legitimacy through common social policies to counterbalance the economic rigours of a single integrated European market and a shared currency. Sustained by a recent shift towards social reformist parties in government viz. France, Germany, Netherlands and the UK, there appears, particularly in the European Parliament, to be a shift towards stronger transnational integration, dependent upon successful realignment of the financial burdens associated with the Common Agricultural Policy (CAP), and a commitment to deal with social exclusion. Several attempts by politicians and policy makers to strengthen European integration, such as the Maastricht Treaty in 1993, or more recently the Lisbon Treaty in 2007, have met with substantial resistance from national states whenever a popular referendum on the EU and constitutional change has been permitted.

Changes associated with postmodernity developed a far more individualised personal philosophy and disengagement from social and political issues. The late twentieth century has seen the rise of more home- and consumption-based consciousness replacing the altruism of post-war reformist welfare collectivism. There has been a diminution of the power of the once all-encompassing city councils. The politics of postmodernity is about distrust, apathy and alienation from conventional political leaders and established political parties. Individualistic lives at work and play found expression in anti-collectivism and a consumer-based privatism which produced a dominant political ideology favouring free markets, personal choice and entrepreneurialism in all areas of public life. In most nation states this has developed into rejection of the old consensus around welfare and increasing emphasis upon self reliance and privatised provision of social benefits.

Growth of a business discourse involving private partnerships, in politics both locally and nationally, has also been matched by growing recognition of the limited sovereignty and autonomy of the nation state. Faced with transnational economic realities and global challenges the power of individual states has diffused towards widening federations which necessarily respond to the demands and constraints imposed by global capitals. Postmodernity has thus been reflected in changes in power structures at all spatial levels and has reconstituted the political realm as much as any other aspect of contemporary life.

Free Time, Leisure and Lifestyle

Leisure in the postmodern city is as diverse, segmented and flexible as the economic, social and political context within which it exists. In a postindustrial world people actively construct their own identities as much from their leisure lifestyles as from any historic relationship with work or workplace. Diversity of leisure lifestyle forms is offered in the cornucopia of a global marketplace communicated transnationally. All that is required to participate is money and the stylistic knowledge or cultural capital necessary for selection and adoption.

Free time in such circumstances is far less normatively circumscribed as barriers between high and low culture, between great art and pop art, between elite and popular forms, become eroded by continuing processes of de-differentiation (Featherstone 1990; McGuigan 1999). The result is an anarchic diversity of fashions and styles, linked only by their propensity to be commodified and sold to the hedonistic identity seekers of the postmodern city.

Flexibility in the labour market subdivides those with too much work from those with part-time or no jobs. One of the paradoxes of contemporary free time is that many experience workplace stress brought on by long hours and intensification of exploitation related to deskilling, downsizing, and increased efficiency targets set for a reduced workforce, while others can find only part-time, temporary, or no work at all. Contrast in lifestyles and intensity of free-time activity is real for those experiencing either end of this postmodern spectrum, too much work or too little, time rich or time poor, both equally unwelcome and destructive of personal and family well being.

Quality of life and leisure experiences assume ever greater significance in this context. Economic activity based on commerce and services is less tied to any specific location and so can afford to indulge individual preferences for an attractive urban or semi-urban environment linked to maximising the satisfaction of the decision makers. Quality of life from a free-time perspective takes a diversity of forms, deriving, perhaps, from picturesque settings, high technology access, cultural opportunities, or active recreational locations. With such diversity urban centres attempt to develop whatever they consider distinctive about their own cores as a marketing device linked to a suitable postmodern ambiance for commercial investment and economic activity.

Leisure Resources

Resources for leisure are both personal and local. Individuals are distinguished in Pierre Bourdieu's (1984) terms by their developed capacities to engage with particular cultural forms. All leisure activities carry the potential to contribute positively or negatively to an individual's social standing or peer status position. Historically, the requisite understanding to enjoy opera or ballet or other classical cultural forms was seen as a social mark of distinction and represented attainment of cultural capital, much as economic capital might be accumulated in other spheres.

Distinction between high and low cultural forms has been successively eroded by the postmodern de-differentiation of historic values and their replacement with relativistic perceptions of worth and status. New forms and traditional practices have been adopted and accepted as worthy in this postmodernist diversity of styles. Old patterns have been partly overthrown though there still remain distinctions within the conspicuous consumption of 'right' clothes, food or gastronomy, in conveying status. Naomi Klein (2000) has fiercely criticised globalization and the insidious power of branding, both in the production and the consumption of goods, by global corporations and transnational conglomerates.

The peer group centrally involved in judgment has shifted. The old bourgeois collectivity has been supplanted and enlarged by a growing service class of business and commerce which has come to dominate the new urban economy. This large group of clerical and commercial workers carries its own marks of distinction in its superior working conditions, levels of higher education, literacy and numeracy, familiarity with information technology, and personal mobility. The new mass of urban employees is far better educated and informed and expects and demands for its leisure higher quality facilities and experiences than did the old mass of industrial workers. The restructured workforce with its superior cultural capital thus places further demands on cities.

Local leisure resources vary from city to city but are judged by an increasingly discerning set of local inhabitants and potential investors. Growing individual cultural capital and increased personal mobility have led to far greater awareness of other places and rival attractions. Demands for a high quality ambiance, transport system or city environment are significant and cities must meet service class expectations regarding a superior quality of life. However, a positive aspect of postmodern diversity for cities is that a wide variety of settings are seen as appropriate and acceptable. Whether these cities possess historic buildings like Amsterdam or Bruges, or new high technology forms like Rotterdam, a picturesque location like Edinburgh, or an air-conditioned architectural capsule like Los Angeles, what matters is that the setting should be of high quality. Such diversity accommodates heritage or the contemporary, the active or passive recreationist, but must represent a wide range of facilities and possibilities for both living and consuming and celebrate whatever is distinctive about the location's unique selling point.

Celebration of diversity can thus take a variety of forms, whether in buildings, environmental enhancement and landscape, or special events drawn from history, sport or the arts. What is essential is that cities develop their potential to the full, particularly those relating to leisure and improving quality of life, to attract and retain the increasingly demanding consumers of their facilities and local residents who have to fund local taxes and civic investments.

Urban Recreation

Postmodern leisure is closely linked to commodification, commercialization and mediatisation. Whether in designer sportswear, fashionable drinks, or couture clothing, the brand or label is all important and comes both to symbolise the product and to give status to the act of consumption. This commodification extends to all aspects of leisure lifestyles and partly explains the growth in leisure shopping and expansive retail facilities. Urban leisure is about being out and about in a cosmopolitan setting and about living consumer life to the full, participating in metropolitan cafe society in a city which never sleeps. The design of themed bars and restaurants, with their full-length windows and open aspects are spaces in which to be seen – a far cry from the frosted glass and men-only backroom and snugs of the Victorian era when more segregated gender relations were taken for granted. Civic campaigns in Britain aimed at developing 24 hour cities point to the importance of urban activity extending daily usage of city centres. Urban leisure in the postmodern era is about providing discerning consumers with experiences which are rewarding and fun. City centres reflect this or are abandoned for places which provide a higher quality of leisure life. But there are issues of the serial reproduction of 'non places' in city centres which remain at night-time and become exclusive, youth dominated, alcohol-fuelled, 'club' cities or in US parlance urban entertainment districts. The top ten pub companies own over half of all UK pubs with 70% of beer sales controlled by only three firms. Leeds provides no exception to the rule, boasting Europe's largest bar and 'beer factories', which offer crucial stopping off spots as local and student youths tour or "steam" through city centres on their preferred pub 'runs'.

Growth in domestic leisure clearly poses threats to vibrant central cities. Europe-wide campaigns to reinvigorate city centres, whether in Manchester, Glasgow, Dublin or Lille, point to a decline in their use or to growing problems of accessibility or attractiveness when faced with improved domestic environments. This era can be seen as representing a tension between the decentralisation of homes, free time and work facilitated by increased use of private cars and the residual modernist features forming the core of most urban areas. The challenge for civic authorities is to reverse the centrifugal trends which have seen the growing suburbanisation of leisure and to generate new dynamism in urban centres which will improve the quality of life for current inhabitants (and electors) and offer potential for discerning business investors.

City Centre Management

A key aspect of urban regeneration in postmodernity has been the fate of city centres. Many in Europe and North America have suffered the effects of decentralisation of employment, shopping, residence and even of entertainment. The suburbanisation implicit in greater private car use has had a deleterious impact on city downtowns. In Europe and North America civic authorities have sought to combat decline in a variety of ways. One of the most common has been the exclusion of through traffic and pedestrianisation of large areas of central cities.

This reclamation of the core for the enjoyment of shopping, strolling, and eating out has often been accompanied by investment in fresh paving, new planting, themed street furniture in the form of decorative seats, lighting, fountains, and a range of stylistic and colourful devices aimed at attracting the crowds. At its best the atmosphere of easy tolerance and relaxed enjoyment seen in London's Covent Garden or around the Pompidou Centre in Paris develops a momentum of its own as the street theatre, both official and impromptu, attracts growing crowds of expectant visitors. Atmosphere is all important, hence the need for traffic exclusion, civic investment, and the need for sensitive policing and surveillance to reduce crime and unsightly litter.

This type of high quality intervention is expensive and a number of partnership schemes have been developed, many drawing on Philadelphia's example in which local businesses have paid extra civic taxes in order to improve the quality of the downtown experience and hence their own profitability. Many British cities, like Leeds, Sheffield and Nottingham, and even small towns, have supported this type of transformative policy by employing special civic task forces to run the city centre. City centre managers have been appointed in many cities to ensure that a wholly positive experience is given. The city core is vital for postmodern investment and prosperity and it must have central priority for any civic regeneration strategy. All the North American innovators like Boston or Baltimore have made a new vibrant retail 'festival' market place central to their city core regeneration, and most successful European developments have similarly used the dynamism of a vibrant working and attractive city centre as a vital feature of a successful city image. The city core *is* the city as far as outsiders are concerned and so cannot be allowed to detract from or inhibit economic potential. Leisure investment in an attractive centre is essential and cost effective.

Green politics and single issue movements heighten resource conflicts around the idea of sustainable cities and their potential adoption by resident and commuter populations.

The Postmodern City Context: a Case Study of Leeds

Pulling into Leeds station these days, even when arriving from London, one cannot avoid being impressed by the glittering sets of residential towers lining the River Aire, dominating all approaches to the city. This transformed skyline of a former provincial backwater, symbolises more than anything the resurgence of the city and its mutation into a postmodern city. What were once commercial and industrial buildings now function as lifestyle and leisure spaces. This steady transformation since the 1960s is the culmination of a series of forces and interventions – economic, political and social – changing forever the central areas of the nineteenth century creation which was modern Leeds.

By the 1970s the economic foundations of the Victorian city were being shaken by pressures of global competition. Within 20 years the central economic pillars, whether in clothing, engineering, printing or brewing, were in decline and struggling as major traditional employers found themselves unable to compete with lower priced imports or labour costs generated outside the UK. The decline of firms like Hunslet Engineering, Burtons tailoring and Waddingtons printers spelt hardship for large numbers of Leeds workers and the city struggled to replace lost jobs and revenue in an era of intense global competition. Only slowly was an economic transformation achieved. As a regional centre Leeds always had a large component of its workforce in the education and health services, and to these were slowly added expansion in financial services, banking call centres, accountancy, legal services and retail headquarters. By the 1990s growth in these sectors had begun to turn the tide of disinvestment and replace lost manufacturing jobs with a postmodern focus on service, clerical and office work. A new set of employers such as First Direct, Asda and Halifax, along with nationally recognised accountancy, business consultancy and legal firms, provided a replacement employment base for the city. The ethos of a transformed economic environment came partly through political support from a city council slowly abandoning its Old Labour traditions and changing its attitude towards urban entrepreneurialism. For a short period in the 1980s there was some attempt to object to Thatcherite central government pressure and adopt what Henry (2001) has termed a Left post-Fordist political strategy for the city. But Leeds, under the controlling political leaderships of Councillors George Mudie and Jon Trickett, soon came to adopt more positive attitudes toward change and in aspects like 24 hour city policy learned from Graham Stringer's Manchester[3] or in cultural industries matters from initiatives like those in Sheffield. Such positive support for changing the economic staples of the central city was assisted by Leeds' pragmatic municipal ethos and its position as the regional centre of the West Yorkshire conurbations. This embrace of urban entrepreneurialism was tightened further when the Labour Party lost its long-term overall control of the council in the early 21st century. In 2009 it is run by a pragmatic political alliance of Conservatives, Liberal Democrats and Greens.[4] In political terms, the local hegemony of parochial industrial Labourism is over. Leeds now looks further afield to provide regional leadership and to become an attractive European city.

Unlike several other nineteenth century centres, particularly Bradford and Birmingham, Leeds always adopted a conservative approach to comprehensive redevelopment of its city centre. This meant that the bulk of its nineteenth century commercial and retail architecture was preserved intact in the late twentieth century and refurbished and reutilised as a positive feature of city image. Classic shopping arcades, like Thornton's and County, remained and were modernised sympathetically. The central retail core, around Lands Lane, derived from Norwich's London Road scheme, had been pedestrianised early, and this 'golden square mile' provided a long-established core of human scale streets and piazzas which could be expanded on and augmented throughout the 1990s.

Leeds was able, like Baltimore in David Harvey's (1989) classic study of postmodernisation, to retain the best of its historic buildings and spaces and to

---

3   For a full discussion of this policy development see Spink, J. and P. Bramham (1999). The myth of the 24-hour city. *Policy and politics. Leisure, Culture and Commerce*. P. Bramham and W. Murphy. Eastbourne, Leisure Studies Association. 65: 139-52.

4   In May 2008 Leeds City Council had 99 members – 22 Conservative, 24 Liberal Democrats, 43 labour, 5 Morley Borough Independent, 3 Green, 1 British National Party, 1 Independent.

complement these with refurbished landmarks like the Corn Exchange, the White Cloth Hall, Leeds Library and the Town Hall with its Victoria Hall. For any aspiring postmodern city the iconography of revitalised Victorian brick and stone added greatly to the attractiveness of the centre for businesses and shoppers. The revitalised Briggate, attracting metropolitan retail magnets like Harvey Nichols department store and the opulence of the Victoria Quarter, ensured the success of the rebranded daytime city centre.

Leeds also facilitated what Harvey (1989) saw as essential for urban regeneration – the opportunity for profitability from commercial exploitation of historically under-valued areas of the central city. This "revalorisation", stemming from a gentrification of land uses and their value, was possible in a swath of central Leeds, alongside the historically polluted river and semi-derelict canal. New riverside and canal side developments, both residential and commercial, replaced a mid-twentieth century industrial landscape of derelict mills and warehouses, scrap yards and builders merchants' premises. Above all, low-value waterside locations were transformed, visually and financially, into attractive propositions for inhabitants and developers alike. The gentrification of central waterside areas provided a stunning corridor of impressive development, embellished by that other essential postmodern icon, a millennium bridge. Thus, large areas of under-utilised and under-valued central city were appropriated as part of a commercially successful postmodern core. Once marginal areas were thus incorporated, the expansion of new employment patterns and the reversal of long-standing population decline became possible. The double benefit was that previously perceived negative areas of the centre became representative of the positive transformation of downtown and served to reinforce the demand for living, working and leisure spaces, centrally and iconically located.

Expanding the centre geographically was paralleled throughout the 1990s by an equally rewarding temporal expansion. Like most provincial cities, traditional modernist central Leeds was largely dead and empty by 6 pm. One of the most significant initiatives taken by the city council, particularly under leadership from Jon Trickett, was to borrow Manchester's innovations in creating a 24 hour city. A 'city that never sleeps' came to be seen as a central aspect of postmodernisation. The expanding service sector of bars, restaurants, discos and clubs, reflected contemporary economic growth and was accompanied by attracting suitably 'beautiful people' parading as consumers and entrepreneurs. Leeds duly expanded its licensed premises and its image as a 'fun city' in which to live, study, work and play (Spink and Bramham 1999). The London media were convinced and the success of the 24 hour city initiatives attracted considerable publicity which in turn reinforced the success of this development. Expansion in service sector employment and higher education participation, all helped fuel the night-time economy boom, essential for the image and attractiveness of a postmodern city. Initial success was consolidated and accordingly the night-time sector has itself helped to contribute largely to recent commercial success in the city.

Thus, Leeds from around 1980 onwards developed an integrated approach to the transition from modernity to postmodernity. It coped with the economic losses of deindustrialisation by diversifying into new areas of commerce and services. It used its Victorian heritage to good effect aesthetically in the transformation of its pedestrianised central retail spaces. It seized upon its under-utilised land resource along the waterside corridor to transform the image of formerly run-down and low-value real estate into a set of flagship and landmark buildings, marking and framing the entrance to the city. It established a reputation for 24 hour activity and cosmopolitan living to revitalise and repopulate its residential core. It managed to reverse retail decline and decentralisation to create an attractive shopping environment of regional scale in a humane cityscape. Above all, it succeeded in reinventing itself to remarkably good effect in a way that few of its regional rivals have achieved. In urban life such success is often reinforced cumulatively by new investment, and the dynamism of postmodern Leeds, once generated, shows no sign of slowing – the real indicator of a successful postmodern city.

Postmodernising Places

Structural economic change has been reflected in the physical reconstruction of the city. Key elements of the serial reproduction that David Harvey (1989) identified as characterising the developing global postmodern city are evident in European cities, such as Leeds. In each there has been considerable investment by the public sector, local urban development corporations and the private sector, in real estate which symbolises the economic shift to clerical, commercial and retail work. This reconstruction has created economic advantage in a number of prime locations in most European cities, typically exemplified by:

- **precinct, plaza and mall** - the extensive refurbishment, pedestrianisation and construction of central shopping facilities and the interlinking historic arcades and streets. This has involved the utilising of themed street furniture in local authority projects partnering local businesses to retain prosperous retail trade in a vibrant 'downtown' core, stretching across the centre from Millennium Square to the railway station and the markets area.
- **gateways and landmarks** - the establishment of substantial office complexes built in a conspicuous 'postmodernist' style of architectural eclecticism involving colour, pastiche, historic detailing and vernacular brick or stone. These complexes symbolise municipal and commercial optimism by occupying major entry point sites like those at the ends of the principal downtown axis, the Headrow, or framing the commercial cores of cities in a swath alongside the river Aire and canal.
- **waterfronts and docklands** - in most cities with postmodern pretensions there has been an explicit policy focus on improving waterside environments along docksides, rivers and canals. These have been transformed from semi-dereliction into prestigious locations for offices, hotels, new museums, waterbus ventures and waterside walkways, aimed at the growing service

classes in work, leisure or tourist modes, particularly transformative in Leeds' case from the state of pre-existent dereliction to prime site status.

- **business, retail and science parks** - the construction at a number of key locations on radial routes outside the central city of estates aiming to diffuse employment and to maximise the advantages of increased car accessibility. This has led to the decentralisation of jobs onto promoted sites, where mobile capital in the form of commercial investment is encouraged to occupy low density postmodernist 'themed parks'; whether White Rose Centre to the south west, Owlcotes to the west, Crown Point to the south or along the ring road to the north of the city centre.

These spatial relocations have been articulated through the operation of a series of urban real estate processes. Local, national and international property development capitals have been engaged in maximising the potentialities latent in the extensive economic restructuring of the city.

Postmodernising Processes

In most contemporary cities there has been extensive *gentrification* as selected and appropriate inner urban areas have been redeveloped and colonised by higher social class groups. Many of the growing numbers of students, clerical and commercial personnel have chosen to occupy former working-class, rented, terraced neighbourhoods like Burley, Kirkstall or Chapel Allerton, since these provide accessible, convenient and affordable central accommodation for housing purchase or rental (Bramham and Spink, 1994). These first steps on the owner-occupation ladder are often facilitated by the presence of large numbers of relatively cheap properties available for purchase and refurbishment in particular areas of cities which could be colonised at that time without too great a danger to the individual or their investment capital. The displaced tenants and former populations of these rising affluent areas were effectively marginalised and obliged to share the spaces of the 'ghetto poor' in the less marketable, often crime-ridden, areas of the inner city and peripheral large 'sink' estates.

This *polarisation* clearly reflects the growing gulf between affluence and squalor in what has been termed the 'dual city' of postmodernity, and came to be represented in Leeds as in many other metropolitan centres like London, Rotterdam or Paris by the 1990s (Byrne 1995). Leeds has its own share of 'inner city' areas like Little London, Sheepscar or Beeston, which suffer high rates of deprivation, unemployment and racial tensions, as well as unregulated and decontrolled leisure, such as under-age drinking, joyriding in stolen cars, drug use, prostitution and a quiet undercurrent of violence and sustained criminality. The marginalised homeless, mentally ill and alcoholic thus still make fleeting appearances amongst the shopping malls, pedestrianised city streets, arcades and green spaces on the edge of prestigious city buildings and thoroughfares.

Gentrification of cheaper property and its enhancement in value is a project pursued by finance and development capitals elsewhere in cities, but on a much more commercial scale. The waterfront transformation mentioned earlier represents a prime example of successful *revalorisation* of recently undervalued areas of real estate which were transformed in function, appearance and value in a way which has proved so profitable in city centres throughout Europe and North America. Along waterfronts displacement of poor quality land uses has ensured that developers have achieved high rates of return as once semi-derelict spaces have been 'revalorised' into office blocks and warehouse conversions for residence.

Investment has often been facilitated, as in Bilbao or Rotterdam, by the entrepreneurialism of the local council working in direct partnership with private investors, through the Leeds City Development Company and by the operation of the Leeds Development Corporation as a central government quango, established in 1988 to co-ordinate and facilitate investment by private capital (disbanded in 1995). Emphasis on these central and public sections of the city has been as *individualistic* and *anti-collectivist* as the trends of the last 20 years in domestic life. The liberalism of New Right central governments has encouraged private initiatives wherever possible and advantaged private agencies over public bodies. Accordingly, private capital has invested and reaped the benefits of revalorisation and has privatised great swaths of land, both waterside and in the city centre.

Privatisation has not only been promoted by government in the UK through sales of property like council houses, school play spaces and hospital buildings, but also in the tendering, contracting or deregulation of service provision and the commercialisation of management and control in once purely public agencies. Within central cities this process has been manifest in the privatising of formerly public spaces, as the enclosure of city streets to form shopping malls has demonstrated. Once public thoroughfares, like Queen Victoria Street, have been subsumed by trendy boutiques, luxuriant plantings, textile hangings and terrazzo flooring to create a postmodern ambience of sophisticated and yet camera-surveyed shopping pleasure within a space controlled commercially, atmospherically and financially, as the Victoria Quarter.

Polarisation and fragmentation of the city thus deepen as surveillance cameras and private security guards add to divisions of the populace into 'insiders' and 'outsiders'; divisions which are at the core of revamped city centres. Those marginal to the 'new' economy, whether through the economics of the labour market, and/or through their location in unfashionable areas of the inner city or the outer estates, are effectively excluded from the 'new' city. In the past the plight of the excluded would have been the central concern for a welfarist local council. Recently, city councils like Leeds, Sheffield and Manchester have pursued policies more varied than the historic social reformist focus on inequality and social justice. This has not disappeared altogether but was recast as Leeds' Corporate Plan for 2005-2008 'to bring the benefits of a prosperous vibrant attractive city to all the people of Leeds'. The major political paradox facing 'postmodern' Leeds is its current discourse of inclusion, a utopian future of participatory citizenship, with little acknowledgement of a major undercurrent based upon the politics of racial difference, inequality and anti-globalisation. For example, in 2006 a British National Party candidate was elected in a ward in South Leeds and, perhaps ironically, Leeds City Council's long-term community planning document, 'Vision for Leeds 2004-2020' is no longer prominently available on the local government internet site, but is only accessed through the Leeds Initiative website. Traditional demands for social justice have literally become diluted by aspirations to become a European City. As a result, the three stated priorities for Leeds are first, *Going up a league as a city* – making Leeds an internationally competitive city, the best place in the country to live, work and learn, with a high quality of life for everyone; secondly, *Narrowing the gap* between the most disadvantaged people and communities and the rest of the city; and third *Developing Leeds' role as the regional capital*, contributing to the national economy as a competitive European city, supporting and supported by a region that is becoming increasingly prosperous. In the recent past, the focus of the new urban left has been much more accommodating or pragmatic when faced with continuing financial constraints from central government. Recent council leaders have seen the need to join with private developers in extensive area redevelopments. This has not been restricted to the redevelopment of decaying areas, but has extended to prime sites for new development around ring roads and in the green belt surrounding cities. This 'new realism' has emphasised the need to project a progressive image of party, local government and the city, and this approach has been incorporated into evolving municipal policies.

Projection of a supportive and attractive image for inward capital investment has been pursued partly through the 'beautification' of city centres and by leisure policies more generally. City image initiatives have taken a variety of forms, designed along the lines of city promotion and civic boosterism detailed elsewhere (Harvey 1989; Law 1993; Spink 1994). Investments in 'hallmark' events in Leeds' case (including a national cycle race), in cultural initiatives (Opera North, West Yorkshire Playhouse, the Phoenix Dance Theatre, the Northern Ballet Theatre, Yorkshire Dance and the Henry Moore Sculpture Centre), and in 'heritage' (e.g. Royal Armouries Museum, Armley Mills, Thackray Medical Museum and also the Corn Exchange refurbishment), all support a postmodern focus on the city as a key component of a fashionable and attractive environment, positively contributing to an appropriately commodified quality of life. The shibboleths of welfarism dissolve and so in the words of Kevin Johnson, Marketing Leeds chief executive launching 'Leeds Live it Love it' in Autumn 2005, "Our brand serves many purposes - it is a promise, a call to action and a statement of pride. Our proposition is neither conditional or specific – it is unconditional. It does not depend of class, ethnicity, background, gender of upbringing. Leeds is a place where you can live it and love it".[5]

Cities have frequently taken particular care with building styles in order to maintain a uniform postmodern 'look' (of eclecticism, echoes of the vernacular,

---

5   See Lee http://www.leedsinitiative.org/.uk Autumn 2005 Newsletter.

pastiche and use of brick, stone or tile); they have invested in new street furniture to integrate a restyled pedestrian centre; they have developed riverside walkways complete with the 'indispensable' postmodernist icon of a suspension bridge; and they have initiated tourism strategies to take advantage of new museums and transformed watersides. In policy documents local authorities adopt this now standard postmodern approach towards city integration being followed by rival urban centres across Europe. The segregated vision of post-war urban planners which separated industries, housing, shops, parks and public buildings is replaced by the 'integrated city' where affluent consumers live, work and relax. This model of the 24-hour 'continental' city has been adopted explicitly by leaders of city councils like Leeds, Manchester, Nottingham and Leicester as a major policy objective and as such symbolises the postmodern approach to the city and to urban politics. Instead of the old Labourist concern for welfare and collective provision, the city of the millennium is to be that of the sophisticated 'European experience'.

The vision of a 24-hour city in which contemporary lifestyles are to be realised is part of the policy both to kick start the economy and to transform the quality of life for residents. It implies a new approach to city planning and development and also to council policies, including leisure, which is much more redolent of the New Urban Left of the 1970s or Right Post-Fordism than the old Labourism of modernity (Henry 2001). Traditional Labourism may persist in other areas of council policy such as 'Social Services' and 'Community Benefits and Rights', but it is the position or relocation of leisure policies we are concerned with here.

A number of processes are operating in contemporary cities. Some are related to individual decisions made by the changing labour force as earlier described, while others are due to the operation of the forces of competition between finance and capital. Sub-urbanisation has been a feature of many urban populations since 1945. Demands to move away from cities perceived as posing threats or problems and to make use of improved transport, both public and private, fuelled the movement outwards to areas giving status and a promise of domestic improvement. This outward migration has within Postmodernity produced a counter flow in some settings which has been welcomed by civic authorities. Some city neighbourhoods have been colonised by students and those of higher social status than the present inhabitants. The newcomers, often the service class workers so representative of economic restructuring, have changed the social geography of the city by 'gentrifying' often run-down neighbourhoods. This movement and the gentrification of old industrial neighbourhoods thus symbolises the commodification, private individualism and entrepreneurialism at the heart of the postmodern city. It created new land and house values and transformed broad sections of city housing.

Similar processes have been operated by corporate capitals. Here, instead of individual houses, whole areas of derelict or relatively under-used land have been developed for offices or retailing and the land value transformed. In some cases public spaces, even city streets, like the 'Victoria Quarter', from Queen Victoria Street in Leeds, have been effectively closed off and privatised. There has been a gain of new urban amenities in the retail mall or office atrium created, but this has been at the cost of some civic accessibility and the possible exclusion of some elements of the urban population. Postmodern urban processes thus replicate those of all aspects of society in that they can lead to colonisation, exclusion and polarisation. In a setting driven by market commodification and a business discourse some, the less affluent in economic or cultural capital, can lose out in the transformation of traditional cities. Similarly, the cities themselves are necessarily engaged in a desperate regional, national, even continental or global competition for scarce resources and inward investment. In an era of footloose industry, mobile capital, flexible telecommunications and the abolition of the old certainties of locational geography the battle between centres becomes ever more intense and uncertain. In such a climate only those cities which develop competitive postmodernist strategies will prosper.

Leisure and Regeneration

Leisure facilities are vital for key decision makers in reinforcing the city's attractiveness. The presence of sailing, golf, top-class theatre or sport, historic architecture or art galleries and parkland can all influence the decision to bring an increasingly educated and discerning workforce to a particular city. Economic regeneration can thus be seen to be linked with designation as European City of Culture, or City of Sport, Dance or the Arts, rather than with possession of fields of coal or ironstone, as in the nineteenth century. The search for a good quality of life has meant a change in economic location for many employees and businesses. In many parts of north-west Europe it is the smaller, often medieval or historic town which has proved more attractive than the old industrial cities. In Britain, places like York, Winchester, Gloucester, Stirling, or Chester have attracted new service employment and economic growth. Urban growth in France and Germany has also followed some of these patterns of smaller decentralisation as quality of life concerns come to determine contemporary investment practices.

What has come to be vital in ensuring economic success is that cities maximise whatever features distinguish them from their competitors and the way in which civic authorities encourage a pleasing and attractive ambience. The focus can be as varied as the cities themselves and be based on the present or on heritage, on woodland or waterside, but there remains a need for civic investment in the individual elements which give each urban centre its unique identity and attraction.

City Image Projection

European cities just like their North American counterparts have come to see the importance of projecting a positive environmental image in order to attract footloose investment. This 'boosterism' may in reality be centred on a slogan like that used effectively by Glasgow in its 'Miles Better' campaign or by Rotterdam as 'City of Commerce', but positive propaganda has proved important both for local populations and for impressing images on outsiders.

In Britain, Glasgow is perhaps the best example of a depressed and declining industrial city attempting to rid itself of an old inhibiting image. In the early 1980s the city began a campaign to improve the city's image for inhabitants and frequent visitors. It developed a campaign to increase civic pride and used a national Garden Festival to disseminate knowledge of historic architecture and the cultural legacy of the city. Civic authorities invested in new art galleries and the performing arts in order to create a new image of the city. Glasgow was selected European 'City of Culture' in 1990 and used that as a basis for increasing national and international tourism. Leisure and cultural forms were thus deployed to transform the city's negative image and to provide the basis for economic transformation led by new investment. Glasgow realised the importance of a positive image in attracting postmodern business.

Other cities have focussed on sport as a way of transforming their image. Bids for particular mega-events or hallmark events have proved useful, if expensive, for Sheffield (1992 World Student Games) and Manchester (2000 Olympic bids and 2002 Commonwealth Games) or Paris (1998 Soccer World Cup) and London (2012 Olympics). It does not seem to matter whether the sport is land or water based, is team or individual, as long as the opportunity to present the host city in a positive light for the international media is accepted and realised. Cities can thus use whatever assets or events they can develop as long as they market these effectively. Negative images of the past can be usefully dispelled and contemporary positive reconstruction publicised.

Leisure Policies in European Cities

Bramham and Henry et al. (1989) have argued elsewhere that it is no easy task to analyse changes in urban leisure and cultural policies but we must focus on four processes

- the position of cities and regions within global processes of capital accumulation and transnational divisions of labour
- the positioning of cities in the context of central and local state relations
- the local relation between the private and public sectors
- local patterns of leisure consumption.

These four major processes result in uneven division of economic gains and losses throughout Europe. Some cities may gain or consolidate their position in both economic and cultural development (e.g. Barcelona, Frankfurt, Milan, Glasgow) whereas others may be confronted by managing urban decline with enduring divisions and marginalisation (Liege, Birmingham, Sheffield, Marseilles, Naples).

Major cities become drawn into what may become increasingly a zero-sum game in competing to attract transnational capital, national government service sector relocations or regional tourism destinations. Deals and partnerships with non-local capital may equally generate hostility and resistance from local disenfranchised community groups. Secondly, all cities have to face up to and deal with demographic changes–an ageing population, changing and 'thinning' of family structures, increased patterns of urban migration, changing patterns of employment, and growing divisions within racial and ethnic minority populations.

If globalisation and Europeanisation are market-led, transnational and post-fordist in direction, new city hierarchies and city networks will emerge. Some cities have a global role to play in economic and financial markets (e.g. London and Frankfurt). Other cities may benefit from their position as a capital of national states (e.g. Rome, Berlin, Paris) or as centres of regional identities (Barcelona, Antwerp, Bilbao), whereas others may detach themselves to take on a European profile (Maastricht, Manchester).

Civic Strategies for Postmodernism

Irrespective of local partisan politics, many major cities have developed a cluster of leisure policies and initiatives which have a Post-fordist character. Right-Postfordism has been strongly influenced by the ideas and the institutions of the New Right in the UK (Henry 2001). These involve a minimalist role for the state, especially in direct provision, and increased role for the private sector, commercialism, entrepreneurialism and new forms of leisure initiatives.

One key change is that leisure is seen less as part of a welfare rationale and portfolio of services and much more an important site for regenerating the economic base of the city by attracting inward investment (regionally, nationally, globally) to provide new service sector work opportunities for local citizens.

Realistic Developments for the Postmodern City

It is clearly the case that cities are in competition with other cities in attracting European capital investment, industrial relocations, new ventures and so on. It is a zero-sum game as transnational companies, central and regional government departments and local businesses have to decide on one location at the expense of others. As has been suggested earlier, there is a tendency towards serial reproduction of low-risk investment in shopping malls, cultural facilities etc. in several cities so that the distinctiveness and identity of places are lost. One shopping mall, or restaurant, pub, and cafe chain, is very much like any other, in any city or region in Europe.

City policy makers need to engage in a realistic SWOT[6] analysis of their city in relation to others in the region or others strategically positioned within other nation states. There needs to be realistic assessment of the focus of leisure policy and appropriate balance struck between high culture, popular cultural forms, sport, recreation and tourism. Rather than embarking afresh on ambitious mega-events with transnational significance, cities have pragmatically developed their own leisure infrastructure and shaped a distinctive leisure identity. Sheffield made the strong and expensive commitment to win the World Student Games in 1991 and now has major infrastructure for hosting sports championships, whether athletics, football, ice hockey, swimming or diving. Investment in mega-events and facilities has often put pressure on the revenue costs of maintaining small scale community-based facilities and in some cities there has been citizen opposition to the costly legacy of specialised facilities and whether large scale facilities actually service the needs of local citizens.

Ecological Acceptability and Postmodernity

Whatever leisure policy developments are embarked upon there are crucial environmental and spatial issues that must be addressed. As in many European cities, Fordist industrial production has resulted in an unwelcome heritage, a legacy of unsustainable developments, water pollution, traffic saturation and deteriorating quality of life for commuters and city inhabitants. Tourist developments and mega-events add to congestion, traffic flows, litter pollution, noise and disruption. Mega-events such as cycle races, beer festivals and pop concerts disrupt completely the daily life of citizens. The restlessness and consumer passions of the tourist disturb the unique tranquillity and ambience of heritage city centre cores.

Civic boosterism can have unforeseen consequences. Investments in retail outlets and city-centre car parking often increase urban congestion. Many cities in the Netherlands in particular have sought a range of measures (e.g. pedestrianisation, bus lanes, traffic calming, integrated subsidised transport systems, cycle routes and secure cycle parks) to separate and isolate historic city centre cores from the noise and pollution of private car traffic flows.

Many cities have sought to celebrate green spaces within the city in urban parks and to develop new green recreation resources within easy access of local citizens, day visitors and tourists. Consequently, European cities are developing large-scale strategies for collecting, sorting and recycling litter, household and industrial waste as well as energy audits in public and private buildings to encourage ecologically sustainable lifestyles. But so far, leisure in the postmodern city, as consumerist and consumptionist as it is, can do little to mitigate its ecological unsustainability.

---

6   SWOT is an acronym for the standard management tool of looking at the Strengths (S), Weaknesses (W), Opportunities (O) and Threats (T) to business organisations and their markets.

To do that would require a fundamental change in attitude as comprehensive as postmodernity itself.

Postmodernity facilitates a renaissance for cities fitting the demands of economic and technological restructuring. The emphasis by mobile capitals on quality of life reinforces the significance of a preserved, leisured and attractive central city environment. Cities need to be aware of the social, cultural and economic processes of the postmodern which provide the context for leisure development. The politics of the postmodern city have been subject to little detailed research. Politicians and policy makers have to construct distinctive and sustainable policies which provide a clear identity for the city, its heritage and cultural identity and meet the diverse needs of local citizens and outside investors alike. Cities which have achieved that fine balance have been the ones which have most successfully transformed themselves into postmodern cities.

References

Bourdieu, P. (1984). *Distinction: a Social Critique of Taste.* London: Routledge.

Bramham, P. et al. (eds.) (1989). *Leisure and Urban Processes.* London: Routledge.

Bramham, P. and Spink, J. (1994). Leisure and the Postmodern City, in *Modernity, postmodernity and lifestyles,* edited by I. Henry, Eastbourne: Leisure Studies Association Publications, No. 48, 83-103.

Castells, M. (1996). *The Information Age.* Oxford: Basil Blackwell.

Coltrane, S. (2004). Research on Household Labor: Modelling and Measuring Social Embeddedness of Routine Family Work. *Journal of Family and Marriage,* 62 (4), 1208-33.

Featherstone, M. (1990). *Postmodernism and Consumer Culture.* London: Sage.

Harvey, D. (1989), *The Condition of Postmodernity: an Enquiry into the Conditions of Cultural Change.* Oxford: Blackwell.

Henry, I. (2001). *The Politics of Leisure Policy.* Basingstoke: Macmillan.

Hoggart, R. (1957). *The Uses of Literacy.* Harmondsworth: Penguin books.

Klein, N. (2000). *No Logo: taking aim at the brand bullies.* London: Flamingo.

Kreitzman, L. (1999). *The 24hr Society.* London: Profile Books.

McGuigan, J. (1999). *Modernity and Postmodern Culture.* Buckingham Philadephia: Open University Press.

Ritzer, G. (2004). *The Globalisation of Nothing.* London, Thousand Oaks, New Dehli: Pine Forge Press, imprint of Sage publications.

Spink, J. and Bramham, P. (1999).The Myth of the 24-hour City, in *Policy and Politics. Leisure, Culture and Commerce*, edited by P.Bramham and W. Murphy. Eastbourne: Leisure Studies Association Publications No. 65, 139-51.

Spink, J. and Bramham, P. (2000). Leeds: Re-imaging the 24 hour European City, in *Leisure Planning in Transitory Societies*, edited by M. Collins. Eastbourne: Leisure Studies Association Publications No. 58,1-10.

Spink, J. (1994). *Leisure and the Environment.* Oxford: Butterworth Heinemann.

Taylor, I. *et al.* (1996). *A Tale of Two Cities: Global Change, Local Feeling and Everyday Life in the North of England – a Study in Manchester and Sheffield.* London: Routledge.

Chapter 3

Cranes Over the City:
The Centre of Leeds, 1980-2008

Janet Douglas

The transformation of Leeds from an allegedly decaying Victorian city into a thriving, vibrant 'core city' has been ardently promoted by the City Council's publicity machine and applauded in a host of articles in the national broadsheets. Undoubtedly over the past twenty years, the city has experienced a period of exceptional growth. Since 1998 over £2.2 billion property development schemes have been undertaken in the centre and a further £1.6 billion developments were under construction in 2008 (Leeds Economic Handbook 2008). 'Cranes over the city' has become a familiar *Leitmotiv* of self-congratulation and one frequent refrain, addressed to those who have not visited Leeds for some time, is 'you wouldn't recognise the place nowadays'. Even some academic commentators seem to have been caught up in this general euphoria but my argument in this chapter is that this 'renaissance' has been greatly exaggerated and even fetishized. There have been both successes and failures and ahistorical approaches have blinded observers to significant continuities.

Cities are never static and each succeeding generation perceives the rate of change as amazingly rapid. For instance, The Leeds Guide of 1806 declared that 'every year has witnessed an increase of buildings having started into existence with a rapidity which constantly afford astonishment in the minds of the occasional visitor' (18-19), whilst an Old Leeds Cropper in a poignant introduction to 'Old Leeds' published in 1868, wrote of 'living to be aggrieved at a too rapid progress, outstripping my wishes, ideas and my affections' (2). Simon Gunn (2000: 231) explains how between 1840s and 1880s cities were remodelled, 'the city-centre was increasingly reserved for business, warehouses, consumption and civic administration, these functions themselves occupying different zones within the centre'. Except for the reference to 'warehouses', this spatial form is one we would still recognise. Some key features of urban governance today would also be familiar to 19th century urban historians: the fragmentation of power, the importance of non-elected institutions, partnerships delivering various services and the necessity for coalitions and networks to realise changing policies. Intercity rivalries are nothing new. In the nineteenth century, cities competed with one another to construct the tallest buildings or the largest town halls. Today the hubris of height is still with us and cities vie with each other to sell the first £1 million penthouse. What is however, very different is the primacy of 'the local' in the 19th century. Businesses were locally owned, members of the Leeds elite, if not originally locally-born, became committed to the town and since at least the eighteenth century there had been a self-conscious identification with the place. According to R.J. Morris (2007), the nineteenth century was the heroic age of municipal corporations as councils responded to market failures. What happens in the twentieth century, is 'the municipal as discontent', caused by rapidly expanding powers of central government, the demise of local industry, bureaucratic failures and growing disenchantment with most forms of authority. The nationalisation of British life and later its internationalisation led W.G. Hoskins, the doyen of local historical studies, to predict that 'by the 20th century, there will be no local history' (Morris 2007).

The champions of the urban renaissance model choose to depict pre-1980 Leeds as a grimy industrial mill town, the eternal Victorian city, and yet it is possible to construct a different narrative. First, although there were factories in Leeds, it was never primarily 'a mill town'; few cities were. As well as the suburban explosion of the 20th century, even in the city-centre there were significant changes. The 1935 Yorkshire Post's Guide to the city proudly proclaimed that:

> The centre of Leeds has been transformed since the War. There is still some old property in Briggate, one of the chief shopping streets, but most of shop fronts have been modernised with the use of striking designs, in which much attention has been paid to the use of Metal, Glass and concealed lighting ... The same comment can be made of Commercial St., Bond St. and, naturally, the big new street The Headrow ... Electric signs have proved increasingly popular, so that both by day and night, the city is a great attraction not only to inhabitants, but those who live in the densely-populated area outside. (Burt and Grady 1994: 209)

The new Headrow, a boulevard designed by Sir Reginald Blomfield in 1925 and modelled on London's Regent St., was driven through an area of crumbling property on the northern edge of the city-centre. One of its major attractions was the Lewis's department store, which was the first building in the city to cost more than a million pounds. Elsewhere there was the new stylish Queen's Hotel (1936-7), celebrated as 'Park Lane comes to Leeds', and the new railway station concourse which Keith Waterhouse describes in his lovingly-written memoir of growing up in Leeds as:

> ... a breathtakingly spacious chamber straight out of Hollywood, its high, pastel-painted concrete arches hung with what I now know were art deco lanterns, its walls lined ... with glazed terracotta, its floor carpeted with ... faience tiles, its back-lit advertising displays for permanent waves and the New Black Magic chocolates framed in bronze in the manner of the displays in New York skyscrapers. (1994: 57)

By 1926, the new (Outer) Ring Road was under construction and in the early 1930s symbolising its commitment to change, the City Council moved out of its old-fashioned Town Hall and into the splendid Civic Hall. The city's great monument to International Modernism, Quarry Hill Flats (1935-41, demolished 1978) was the pride of the city, built on the site of the city's most notorious 19th century slums. Modelled on Karl Marx Hof in Vienna, it became the best-known Leeds building in the world, continuing to attract international visitors well into the 1960s (Ravetz 1974).

After the Second World War, despite the decline in manufacturing, the city remained relatively prosperous, though central government restrictions meant there was little new building in the city-centre in the 1950s. Even in these years of austerity, R.J Morris recalls that 'The place was full of noise, colour, light, people … it was exciting'. (2008: 4). In 1962, the pioneering Merrion Centre was opened, for a few days Britain's only shopping precinct with shops on two levels, offices, a cinema, nightclub, dance hall, moving pavements and a large multi-storey car park. Two years later the first stretch of the Inner-Ring Road was completed, with its cavern-like tunnels and flyovers: 'it was exciting, terrifying and barbaric … it was clear that something new was happening' (Morris 2008: 5). What followed was a frenzy of construction, Victorian buildings were torn down and concrete office blocks disfigured a number of city-centre streets. In 1970, the first streets were pedestrianised, new shopping centres were built and there were ambitious plans for a high-level walkway linking the eastern and western sectors of the city. Project Leeds, a joint initiative between the City Council and the Chamber of Commerce in 1971 sought to promote Leeds as a dynamic city:

> It is a far cry from the old city spawned by the Industrial Revolution to the modern go-ahead, and continually expanding Leeds of today. Some of the old scars left by that ruthless, thrusting age are still in evidence, but the rejuvenation of the city is taking place on a grand and imaginative scale and at a truly staggering pace. New buildings born of new concepts are pushing their white rectangular columns into the sky … Everywhere there are signs of improvements and progress. Leeds is surging forward into the Seventies. (Leeds City Council 1971: 3-5)

'Leeds: Motorway City of the North' may not have the same resonance today as in the 1970s but it stems from a similar mindset that conjured up more recent slogans such as, 'Leeds: the UK's Favourite City'. This early exercise in civic boosterism even ensured that 'Leeds: Motorway City of the North' was franked on every letter which was posted in Leeds. The architecture of the 1960s and 70s may be unpopular today but it stands as testimony to a modernising spirit which was changing the face of the city.

The OPEC oil price increases of 1973 and the 'Great Inflation' that followed marked the end of the long post-war boom and accelerated trends which had long been apparent in the national and local economy, extinguishing the dynamic of urban transformation. Its political consequences were far-reaching: the end of the social democratic era, the discrediting of the state in both its national and local forms, the burgeoning hegemony of neo-liberalism and the decline of a collective ethos in favour of a rampant individualism. For local authorities these changes were accompanied by the anti-municipal ideology of the Thatcher governments. The restructuring of Leeds that followed in the 1990s, formulated on the twin doctrines of partnership and Europeanisation, was underpinned by a new entrepreneurialism on the part of a City Council which no longer had the capacity to govern alone. Three pivotal developments ushered in this period of transformation: Quarry House, the largest building ever built in the city, was opened in 1993 to house over a thousand Department of Health and Social Security employees transferred from London to Leeds; the city's successful bid for £42.5m Royal Armouries Museum designed to display part of the national collection formerly housed in the Tower of London and in 1996 the opening of Harvey Nichols' first department store outside of London – an icon of retailing made recognisable to people in Leeds through the popularity of the BBC sit-com, 'Absolutely Fabulous'.

To provide the context to the re-invention of the city-centre, the chapter begins by examining the changes in local economy and city governance. The regenerating strategies pursued by the new corporate city form the centrepiece of the chapter which includes three case studies, selected to show links between the economic, political and cultural. The final section looks not at successes but the failures ... projects that miscarried and those parts of the city-centre that have proved resistant to the regenerating impulse. Although it is difficult to predict the long-term consequences of the present banking crisis, already [in April 2009] nine months into the 'credit crunch', many of the ideological assumptions and political discourses that underpinned local regeneration strategies are now in retreat.

The Economic Turnaround 1990-2008

The profound changes in the local economy reflect those taking place nationally and internationally: de-industrialisation, the transnationalisation of firms and the process of tertialisation. Cities have had to adapt to these trends, though how far local policy makers, however mindful they may be of the need for change, can engineer positive results is open to debate. For example, when First Direct opened the UK's first call centre in Leeds in 1988, they explained their choice of location in terms of a large pool of relatively low-cost labour, the city's transport links and other factors such as the accent of local people (Tickell 1996:111). Nevertheless policy makers in Leeds came to believe that they had no alternative but to develop strategies to court foot-loose capital or else watch their city decline into a provincial backwater.

The withering away of local industry has been a long, drawn-out process. Between 1851 and 1911, the numbers employed in manufacturing fell from 74 percent to 64 percent; by 1951 just less than half the working population was engaged in manufacturing. Like other northern industrial cities, Leeds had failed to attract new industries and came now to rely on the tertiary sector. As early as 1911, a quarter of the work force were engaged in the service sector and forty years later this figure had increased to a third – a figure which would have been higher if white collar workers returned in the manufacturing sector by the Census were redistributed (Rimmer 1967:158-178). In the decade between 1981 and 1991 manufacturing dramatically declined from 31 percent to 22 percent and today only ten percent of the work force is engaged in industry.

There was little multi-national presence in Leeds before the 1980s but today what remains of manufacturing is no longer in local hands: in 1990, the world famous manufacturer of surgical equipment, Charles Thackray, was bought by Boehringer Mannheim of the USA, the remnants of the once-proud engineering industry are now owned by German and Dutch firms, in 1994 Waddington's sold their Games Division to the American Hasbro, and the firm has now disappeared under the umbrella of Communsis PLC. It is highly symbolic that the site of its main factory is now occupied by First Direct. That titan of the local industrial scene, Tetley's ('Yorkshire Men are Tetley's BitterMen') became part of the Carlsberg empire in 1997 and, at the end of 2008, the parent company announced the closure of its Leeds plant due to falling demand.

The devastating decline of manufacturing has, at least in terms of job numbers, been balanced by expansion in three tertiary sectors. Between 1991-6 for example, there was a 22 percent increase in Financial and Business Services (FBS), a ten percent growth in public administration and six percent increase in Distribution, Retailing and Leisure (Unsworth and Stillwell 2004:169). By 2002 almost 70 percent of jobs in Leeds were to be found in the service sector. The most dramatic increases were in FBS: by the end of the century a staggering 92,000 people worked in FBS and today Leeds is the largest centre for financial services outside of London, providing between a quarter to a third of all jobs in the city. Accommodating this sector means a huge demand for office space. According to the City Council between 1994-2004, £569m of office property investment was recorded in the city, 63 percent of it in the central core and frequently in tall buildings. Leeds now ranks 15th in the world league table for office rental values (Leeds Economic Handbook 2008), and the local press rubbed its hands with glee when it was announced that rents were 40 percent higher than some parts of Manhattan.

An essential element in securing Leeds' economic success has been to make the city attractive to consumers. The city-centre now boasts five miles of shopping streets and 700,000 people shop there every week. Retailing forms only a part of the boom in consumption. By the end of the 20th century there were 105 bars, 62 restaurants and 25 nightclubs in the city, the number of hotels increased between 7 in 1989 with 850 rooms to 23 by 2007 (with 3,387 beds) and a further ten hotels are in the pipeline. In part this growth in hotel accommodation is fuelled by an increase in tourism – that Leeds should have a tourist industry would have been unthinkable thirty years ago.

The Growth of Urban Governance 1990-2008

Sir Charles Wilson, Conservative Leader of the City Council between 1907 and 1928 once described the council as 'the Do-it-all Corporation' (Leeds Tercentenary Handbook 1926: 68-71), but its activities never extended to the direct encouragement of outside investment in the city or in private-sector job creation. The task of the local authority was to deliver a range of services to the city's residents and, irrespective of political party, this was achieved by a political system characterised by strong political leaders often in alliance with powerful chief officers. The shift from urban managerialism to urban entrepreneurship in the 1980s saw the emergence of new organisational forms based on public/private partnerships that produced less clear-cut, quasi-federal arrangements wherein power became camouflaged and difficult to locate.

From 1979 until 2004, the city had a Labour administration and, much to its fury, in 1987 the central government imposed an Urban Development Corporation on the city. Believing that Labour Councils were too hidebound to spearhead regeneration programmes, the third Thatcher government granted Development Corporations sweeping planning powers to fast track development and budgets to prime the pump of private-sector investment. Bypassing the local authority, the Leeds Development Corporation (LDC) was given control over great swathes of land in the city-centre along the waterfront, encroaching into Holbeck, Hunslet and along the Aire Valley. When the LDC was finally wound up in 1995, they had spent £72m on regeneration projects and one of the effects of this alternative centre of power had been to instil a more pro-active attitude to regeneration and the need for co-operation with the private sector amongst the Council and its officers.

Learning to work together was not always straightforward. In the heady years of the radical Alternative Economic Strategy, some members of the Labour Group, not to speak of the wider party membership, believed that partnership projects were akin to 'sleeping with the enemy', in the business world many perceived the Council as antipathetic to their interests, an obstacle rather than a possible partner. A telling illustration of this is recounted by Adam Tickell: after the 1989 Census of Employment revealed the growing importance of FBS, the Council decided to invite financiers and lawyers to a meeting and dinner at the Town Hall to discuss areas of common interest. Attendance was thin and those who did attend were largely hostile, making it clear to the Council that FBS could best flourish by the local authority leaving them well alone (1996: 113).

Gradually entrenched attitudes began to erode. Within ruling Labour circles there was a growing realisation that there were simply no alternatives to co-operation and partnership if lost jobs were to be replaced and revenue streams maintained. The Chamber of Commerce was also coming 'on message' in response to central government promptings and fears for its own survival. It even considered plans for devising its own economic strategy, and thus usurping a role which many Labour politicians and their officers felt rightfully belonged to the City Council. At this critical juncture, in 1989, the election of Jon Trickett as Labour Leader galvanised this new spirit of partnership and a year later, eleven years before it became mandatory, he launched the Leeds Initiative as a way of opening up a dialogue with the main economic interest groups in the city.

The membership of the Leeds Initiative is revealing of its goals as an economic development forum: the agency is chaired by the Leader of the Council whilst its Vice-Chairman is the President of the Chamber of Commerce, other members are drawn from the local offices of the Departments of the Environment and Industry and Trade, the city's two universities, Yorkshire newspapers, the regional TUC, the Urban Development Corporation and the Leeds Training and Enterprise Board. Its six objectives, unchanged nearly twenty years later, were to:

- Promote the city as a major European centre
- Ensure the economic vitality of the city
- Create an integrated transport system
- Enhance the environment of the whole city
- Improve the quality and visual appeal of the city
- Develop the city as an attractive centre for visitors.

Around this business-friendly agenda and under the umbrella of the Leeds Initiative, a plethora of Strategy and Development Partnerships, Delivery Partnerships and Local Partnerships have developed, at least twenty-four in total, covering everything from climate change to culture.

The election of a Labour Government in 1997 opened up opportunities for a broader more inclusive approach to urban regeneration which if little else, had the effect of bringing representatives from the voluntary sector into the Leeds Initiative's fold. However in 'From Corporate City to Citizens City?', Haughton and While (1999) tentatively argued that this more holistic approach entailed a moving away from a narrow pre-occupation with economic development towards a broader agenda that engaged with social and environmental matters . The city's response to these changes was embodied in 'Vision for Leeds' published under the auspices of the Leeds Initiative in 1997, the draft of which had been widely distributed in the city as a part of an extensive consultation exercise. A second 'Vision for Leeds' followed in 2004, highlighting the need 'to go up a league, as a city – making Leeds an internationally competitive city', and 'narrowing the gap between the most disadvantaged people and communities and the rest of the city'. It is difficult today to pick up a Leeds Council or Leeds Initiative publication that does not repeat these twin mantras and it follows that the Leeds Initiative now has a Going Up A League Board and a Narrowing the Gap Board, both responsible to the Initiative's Executive Board. Labour's loss of power in the city in 2004, replaced by 'a rainbow coalition' of the Conservative, Liberal Democratic and Green Parties, has so far had little impact on the prevailing policy agenda, continuing a tradition of consensus that was established in Leeds politics after the Second World War (Hartley 1980: 439).

Quite how this quasi-federal partnership structure holds together is problematic. Despite the rhetoric of coalitions for growth and opportunities for networking, one can sympathise with the bewilderment of Nigel McClea, Head of the Leeds Office of the legal firm Pinset Masons, when he commented that

> Getting to know Leeds is ... not easy. There is no manual to describe who controls the city and how the city operates. Leeds is probably controlled by a hundred people. (2008: 185)

We have no studies of the extent to which the Council's Executive Board is the ringmaster of this unwieldy conglomeration of agencies, no research into cross-cutting membership of partnership organisations, let alone empirical investigations of informal networks based on political party, faith groups, common backgrounds and neighbourhoods and clubs and associations that promote common interests, cohesion and the trust which are hallmarks of a power elite. More than ever local policy-making is also dependent on a whole array of external power holders, state agencies, development quangos and funding bodies such as English Partnerships, the Heritage Lottery Fund, the Arts Council and Sport UK. All add to the complexity of the local political landscape. No amount of public consultation and notions of inclusion can camouflage the 'democratic deficit' which lies at the heart of these arrangements that have minimised public scrutiny and accountability, resulting in widespread feelings of powerlessness.

Leeds Re-invented?

During the 1980s all over North America and Western Europe, cities grappled with consequences of de-industrialisation and neo-liberal agendas. Salvation was sought for their troubled cities in programmes of civic renewal: encouraging office building particularly for the financial sector, city-centre living, new retailing opportunities and a proliferation of restaurants, bars, hotels and cultural projects that came to signify the vibrancy of city life. These initiatives came as an interconnected package, each element serving to reinforce the potential success of the others. In the local context, Jon Trickett is usually credited with bringing the vision of a European-styled urbanity to Leeds. As we saw earlier, it was Trickett who spearheaded the setting up of the Leeds Initiative in 1990, and a year later he spoke at a 'Bringing the City Alive' conference in Leeds unveiling a vision of more pedestrianisation, a return to city-centre living and giving Leeds 'a continental feel'. Many of these themes re-appeared in Trickett's speech to the first 'Twenty Four Hour Conference' held in Manchester in 1993 but on this occasion he also emphasised the importance of city-centre shopping as an experience shared by all residents of the city, 'the role of retailing as an animateur is crucial'. He went on to explain to his audience that the City Council was developing what he called 'an events strategy' and cited the examples of the Christmas Lights Programme, the three-week 'Rhythms of the City' Programme and the Valentine Fair. The latter first introduced in 1992, was described in hugely romantic terms:

> ... for a week at least, the whole character of the City Centre is transformed from 6pm to midnight. Young lovers mingle with grandmas and granddads as Europe's largest Ferris wheel spins against a backdrop of Leeds Town Hall. The surrounding bars, cafes and restaurants exude a warm carnival atmosphere. And in an abrupt change of key, he added 'after an initial pump priming, it costs the Council Tax payer not a penny'. (1994: 9-11)

To realise this dream of urban conviviality which was so often expressed in terms of becoming 'the Barcelona of the North', licensing regulations for pubs and bars were relaxed. Proprietors were actively encouraged to place tables and chairs on pavements, a French boules court and chess tables appeared in one of the squares in the business district and seven-day shopping in Leeds was launched with what for some was an implausible slogan, 'Funday, Sunday', all a far cry from the prim Puritanism of Old Labour (Crosland 1956: 354-7). The vehicle for many of these innovations was the City Centre Initiative set up in 1993 under the auspices of the Leeds Initiative, and a new civic task force, the City Management Team, which was put in place to ensure that 'conviviality' did not get out of hand. Both were to become key players in pushing the corporate-led redevelopment of the city-centre. Jon Trickett left Leeds to become MP for Hemsworth, West Yorkshire in 1996, but the partnership structures and philosophy he left behind have been accepted as the new commonsense for urban policy makers.

Office building may not be the most glamorous dimension of Trickett's legacy but it reflects the success of the Leeds Initiative and the Leeds Finance Initiative established in 1993.Cities require an economic rationale and the success of attracting FBS firms to Leeds has been the key driver for many of 'the cranes over Leeds'. Between 1998 and 2007 there have been 398 office developments each valued at a million pounds or more, excluding 'mixed-use schemes' which since 2000 have become increasing popular amongst developers (Leeds Economic Handbook 2008). The most spectacular example of the latter is Bridgewater Place (2004-7), locally known as the 'the Dalek', built by St James's Securities and K.W. Linfoot PLC. The tallest building in Yorkshire and costing £85m, the lower ten floors are let to the law firm, Eversheds and the accountants Ernst and Young and accommodate just short of a thousand employees. Above are twelve floors of apartments plus a café bar, restaurant and retail units (the original idea for the inclusion of a hotel was dropped in favour of more office space). The location of 'the Dalek' south of the River Aire on Water Lane is just outside what was the traditional prime office core of the city along Albion St and the Park Estate; according to King Sturge, the international office consultants much of this stock is out of date and/ or listed and therefore unsuited to large-scale modern office development (Leeds Economic Handbook 2008). Although there remains a high demand for postal code LS1 sites, office development is now expanding north of the Merrion Shopping Centre and westwards along Wellington St and Whitehall Rd.

As well as new offices, the city-centre is also awash with new residential accommodation now referred to as 'apartments' or, more often than not, 'luxury apartments', rather than the more old-fashioned term, 'flats'. According to the 1991 census, only 900 people lived in the centre, by the time of the next census, this figure had increased to 5,050 and by late in 2007, there were 14,000 residents living in 372 apartment blocks. If all the current schemes under consideration in 2007 are to be built (and that is now a big 'if'), the number of people living in the city-centre will have reached 29,000 by 2011(Leeds Economic Handbook 2008). Although when the notion of city-centre living first appeared it was part of the Europeanisation project aimed at enticing local young married couples and retired people back from the suburbs, rapidly these new residential developments have become the preserve of a single social group, the young professionals closely linked with the expansion of Leeds as a financial and legal centre.

One of the consequences of this homogenisation is that the average size of units is falling (they are frequently referred to as 'rabbit hutches') as it is expected that this affluent social group will spend much of their free time outside the home in the city's restaurants and bars. A study undertaken by Rachael Unsworth in 2005 on behalf of K.W. Linfoot PLC, a Leeds firm that has pioneered city living, revealed that about half the apartments were owner-occupied whilst the rest, and an increasing proportion of the new developments, has been financed on a buy-to-let basis by investment consortia. In some of the recent developments owner occupation has dropped to five percent. The vast majority of properties belonging to speculative firms in places like Dublin, Dubai and Australia who have gambled on a profit from well-paid tenants working in the financial services (BBC InsideOut 2006). The question of how many flats are unoccupied is a tricky one; it has been claimed that more than one in ten flats is empty and that over 400 have been unoccupied for more than 12 months (www.bbc.co.uk). An Occupier Survey conducted as part of Unsworth's study showed that 60 percent of the residents were under the age of 30, 64 percent earned over £35,000 and the main reasons they gave for living in the centre were proximity to work and 'lifestyle'. Perhaps the most startling statistic is that 60 percent believed that they would only stay in the city-centre for two years. Another indicator of the geographical mobility of this group is that 57 percent of the respondents indicated that at least one of their last two addresses had been outside of West Yorkshire, and half had moved up to Leeds from the South-East of England. City Living does have some positive aspects to it, despite the poor quality of these apartment blocks; it has livened up some semi-derelict parts of the city, catered for the increasing numbers of single-person households in the city and lowered the need to travel by car (Unsworth 2005). However what this transient, homogeneous, 'alien' population brings to the city other than its purchasing power is questionable: they certainly do not make any connection with the working-class populations living close by and they probably have little attachment to Leeds. They consume some of the things that the city has to offer but what appears to be lacking is any sense of local citizenship and the responsibilities this brings.

Since the mid-18th century Leeds has been a service centre for the West Yorkshire region and shopping has long been one of its attractions. Victorian prosperity was to convert Briggate from a street where people lived to a street where people shopped. Arcades were built in the second half of the century to expand shopping opportunities, one side of the old Boar Lane was demolished in the 1860s to make way for fashionable shops – the Grand Pygmalion, Leeds' first department store was opened here in 1884.The most spectacular retailing development was the Leeds Estates Company development, now known as the Victoria Quarter. Between 1897 and 1900, in two new streets, Queen Victoria Street and King Edward Street and two new arcades, the County and Cross Arcades, there were 200 new shops, numerous cafes and restaurants and a new theatre, all designed by Frank Matcham. However after the First World War, the ebullience of the central retailing area began to fade, the new suburbs and council estates of the inter-war period were accompanied by their own shopping parades, and later, in the 1960s, the arrival of the out-of-town shopping centres posed serious threats to the viability of city-centre shopping.

Unlike many other towns and cities, Leeds Council and its planners, supported by the local Chamber of Commerce, recognised this threat and largely refused planning permission for any large shopping malls on the edge of the city. As already mentioned, Jon Trickett, in what was actually a continuation of council policy, argued that out-of-town shopping would destroy the vibrancy of the city-centre, and interestingly believed that such developments were socially divisive, 'segregating the shopping of the well-off from that of the poor and car-less' (Trickett 1994: 10). Retailing thus became another arm of the Leeds Initiative's regeneration strategy: whereas two thirds of the shopping space opened in Leeds between 1987 and 1990 was out of town, between 1992 and 2001 277 million pounds was invested in shopping and leisure facilities in the city-centre (Unsworth and Stillwell 2004: 254). Cashing in on increasing national prosperity and cheap credit in the period 1998-2007, 95 retailing projects valued at over a million pounds have been constructed, with another 30 schemes in the planning system. City-centre shopping is largely focused on pedestrianised streets or else in indoor shopping locations – the latter include the Merrion Centre (1962-4), the Leeds Shopping Plaza (1974), the St John's Centre (1985), the Headrow Centre (1989), the Victoria Quarter (1989 – 1990) and the Light (2001). In the late 1980s, as part of their modernising agenda, the city had planned another development that involved selling Kirkgate Market and the surrounding area to the Dutch property developers, MAB, a scheme defeated at public inquiry by a coalition of the Market Traders Association and amenity groups in the city.

Of the successful projects, only the Victoria Quarter (1989-90) really contributes to any notion of Leeds as 'the Knightsbridge of the North'. Elsewhere we have the usual array of middle-of-the road or cheap multiples that can be found in most towns of any size – extraordinarily high rents have long squeezed out the independent retailers who bring variety to city-centre shopping which partly explains the fury in 2008 when the leaseholders, Zurich Financial Services, evicted small, and often young, traders from the Corn Exchange to be replaced by an up-market food hall.

Despite these efforts, the city's position in the national retail hierarchy slipped from third place in 2003 to sixth by 2006. According to Phil Crabtree, the city's Chief Planning Officer, it might drop to 15th by 2012, hardly what is meant by 'going up a league' (2008: 21-2) Two new developments are in Crabtree's opinion vital to ending this decline: first, the remodelling of the Leeds Shopping Plaza, the third reconfiguration of this mall, by Land Securities and Caddick Developments (an exciting new roof has been designed by the Catalan architect, Eric Miralles of the Barcelona studio of EMBT), and, second, the entirely new Eastgate/Harewood Shopping Complex which is intended to increase the retailing area of the centre by 50 percent and which has been fiercely opposed by the city's civic societies, and perhaps not surprisingly by Land Securities. Much of the site was cleared decades ago but the new development would demolish most of the remaining buildings including the Lyons Works, a former tailoring factory, which has been developing over the past decade into an embryonic China Town with no help from the City Council. The £700m scheme by Town Centre Securities and Hammerson's, an international developer specialising in retail property, has found a flagship tenant in John Lewis's but a second anchor has proved more difficult: at first it seemed that a Selfridges store might be opened and after their withdrawal, Waitrose which currently has no presence in Leeds, showed interest and currently the developers are dependent on Marks and Spencer opening their biggest store outside of London. The re-construction of the Shopping Plaza is well underway and will probably be completed, but as John Lewis has indicated that they are not prepared to open any new stores until at least 2015, the future for the Eastgate/Harewood Centre looks bleak. After fifteen years of exceptional boom, all the evidence today suggests that consumption as a way of life is faltering. Even Harvey Nichols announced in March 2009 that they had experienced a 40% drop in sales in 2008.

People have always worked and shopped in the city but increasingly in the 20th century, suburban living and television meant that at 6pm almost everyone went home, leaving the streets deserted. 'The City-centre' according to Trickett, 'is a city's impromptu theatre and should be open to all its actors 24 hours a day' (Trickett 1994: 11); 'the cappuccino lifestyle', however, with its mix of cafes, bars and restaurants has turned into the nightmare of 'binge drinking'. The city-centre particularly between 10pm and 4am has become an urban playground with large numbers of young people (100,000 at the weekends) drinking and 'clubbing'. So intent are they on pleasure-seeking that they seem oblivious to the effects of their behaviour both on themselves, and other members of the public they encounter, be they individuals returning home after a visit to the theatre or the cinema, or the unfortunate employees of the bars, clubs and taxi firms who service their excessive hedonism. Park Row, once referred by The Builder as 'the Pall Mall of Leeds' (1896) because of its large number of banks and insurance offices, along with other areas, have been given over to 'vertical drinking' venues owned by national chains of breweries and entertainment corporations.

Culture is no longer regarded as the superstructural outgrowth of a buoyant economic base but an economic resource in itself that can help drive economic growth. Despite the problems of defining 'culture', it is argued that culture can create jobs and generate income; culture can put a city on the tourist map and, more importantly, in a context of inter-locality rivalry, it is believed that the aesthetic ambience of a place positively secures firms' locational decisions. Cultural capital is converted into real capital (Strange 1996: 135). Leeds was not, however, the cultural wilderness that it was made out to be: it had the usual stock of cultural institutions, the Grand Theatre, an art gallery, museums, a season of classical concerts in the Town Hall and a world famous Piano Competition established in 1963. In 1970 the West Yorkshire Playhouse was established, though the fact that it remained in rented university accommodation for the first twenty years of its life hardly suggests that the theatre was a major priority for the City Council. 'It is part of the wound of cities like Leeds', wrote Brian Thompson, 'to ignore this more positive definition of its cultural stock and pine instead for the great lack of this, that or the other that is reportedly ten a penny in the great metropolis' (1971:119). More amazingly, in 1978 Leeds became the home of English National Opera North (now simply Opera North), the only English city outside of London to have its own opera company.

As mentioned earlier, in the early 1990s the Urban Development Corporation and the Leeds Initiative were very anxious to win the bid for the Royal Armouries Museum but perhaps the Museum's difficulties in attracting visitors before the Labour Government abolished entrance charges in 2001 deterred the governing authorities from pursuing other flagship cultural developments. Instead there were a series of smaller scale, ad hoc interventions: free concerts in the park, financial support for the Phoenix Dance Company and the West Indian Carnival and street level entertainments in the city-centre. Compared with other British cities, Leeds was slow to jump on the cultural bandwagon but since 2002, £180m has been invested in the cultural infrastructure of the city. A Cultural Partnership under the umbrella of the Leeds Initiative finally appeared in 2002 though the cultural strategy document of the same year is a particularly anodyne publication full of glossy photographs of people having fun.

The collapsing of cultural hierarchies associated with post-modernity has, we are told, obliterated the old distinctions between art ('official culture') and popular culture, leaving local authorities somewhat adrift on questions of value. In terms of its commitment to continental lifestyles and the distinctiveness of Leeds, one might have expected the city to exploit Opera North as a unique selling point but British perceptions of opera as elitist have muted any such responses, except for the free annual Opera in the Park which attracts tens of thousands of people who combine picnic opportunities with a soundtrack of popular arias. The need to generate large audiences and a provincial/Labourist suspicion of 'high art' have combined to favour a cultural populism which finds its perfect expression in the Leeds Festival which each August since 1999, has organised a weekend of popular music concerts. Now in collaboration with the Reading Festival, it attracts 70,000 visitors from all over the country and tickets sell out in a matter of a few hours. Festival mania marks out the city's cultural calendar: there is the Christmas Festival, a Film Festival, a Food Festival, Festivals of Dance, the annual Breeze International Youth Festival, Waterfront Festivals and so on. Every two years since 2004, FuseLeeds showcases cutting-edge contemporary music in a range of musical genres. This concentration on culture has produced a greater investment in the city's cultural buildings. With the help of a cocktail of funding, Leeds has a new theatre (The Carriage Works), both the Grand Theatre and the City Varieties are undergoing extensive renovation as has the City Art Gallery. A new City Museum opened in September 2009 had 100,000 visitors by the end of 2008. Next year should see the opening of the Northern Ballet Theatre at Quarry Hill (40), now designated the city's cultural quarter and already the home of the Yorkshire Playhouse, the College of Music and BBC North. It is often argued that such programmes are as much about economic development as creativity and that they serve the interests of the influx of young professionals and managers rather than local people, particularly those whose spending power is not high. This in the case of Leeds seems unfair. The foray into the world of dance springs from the genesis of the Phoenix Dance Company in Chapeltown, one of the most deprived areas of the city. This opened up a whole new young and black audience for contemporary ballet. Although this is not the place to discuss community-based arts projects in the city, some of the initiatives mentioned above are free, or at least when the local authority has some say in ticket-pricing, charges are kept to affordable levels, certainly far lower than going to watch the less than 'up a league' Leeds United at Elland Road.

Building the Re-invented City

In his 'Buildings of England: the West Riding' (1959), Nikolaus Pevsner wrote that in Leeds,

> the centre is not without its merits, and this is entirely due to the enlightened planning policy of its corporation'. Blessed with 'a certain orderliness .... Its architecture up to date is not what it might have been but it is unified and its effect on the pattern of the centre cannot be praised too highly. (1959:307)

Ten years later when he visited the city again to prepare the second edition, he found no reason to revise his earlier opinion, except for disapproving of the three tall buildings that had spoilt the appearance of City Square. Not that Pevsner disapproved of tall buildings in principle. As a supporter of the international modernism, for instance, he admired the 1953 Central Colleges, now the Civic Campus of Leeds Metropolitan University. But Pevsner's views were very much out of step with popular opinion: where he saw functional simplicity and abstract form, the general public only perceived the 'brutalism' of the modern movement. There were, however, impressive modernist buildings in Leeds such as the inverted ziggurat of the Bank of England (1969-71) and monumental Brunswick buildings (1973-8)) which if they survive, will be regarded by future generations as significant additions to the built environment. In this subjective context, we need to remember that the Civic Hall was once derided as representing the appalling taste of provincial councillors; now it is regarded as a rather handsome building. By the late 1970s popular anti-modernism was joined by a paradigm shift from within the architectural profession itself: the post-modernist movement with its emphasis on hybridity and eclecticism, historical quotation, sensation and playfulness – some would say 'style over substance' – has become the dominant architectural discourse. Currently a whole raft of organisations and policies both national and local has legitimised this architectural turn. The Commission for Architecture and the Built Environment (CABE) was established in 1999 to influence and inspire decision-making about the built environment and the Urban Task Force Report (2000), associated with the architect Richard Rogers, has been enormously influential; local vehicles for change include the Leeds Architectural and Design Initiative (1994), the Vision for Leeds documents of 1999 and 2004 and the City-centre Urban Design Strategy (2000). The Renaissance Leeds Partnership between the City Council, Yorkshire Forward, English Partnerships and the Leeds Initiative has as its mission 'city shaping' in order to maximise public and private confidence and investment in the re-generation of the city (Wainwright 2009: 11). Over the past ten years or so, new developments under these auspices have largely been undertaken by 'the Big Five': K.W. Linfoot PLC, Town Centre Securities, Landmark Developers, St James Securities and Oakgate, all locally-based firms, and until recently the local architectural firm of Carey Jones have been responsible for the design of most large projects.

Despite criticisms of post-war zoning regulations, Pevsner's 'orderliness', and post-modern notions of variable space, functional mixing and urban surprises, the blueprint for the built environment has retained some spatial specificity now disguised by the new nomenclature of 'quarters'. Thus we find in the north of the city-centre the Civic and Education Quarter, in the east, the Quarry Hill Cultural Quarter whilst the core of the centre is earmarked for retailing. The west is the Financial and Business Quarter, and City Living is focussed to the south and particularly along the Leeds Waterfront. Despite the urgings of the Leeds Civic Trust from the early1970s ('Leeds on Aire'), it was not until 1985 that the City Council began to think seriously about the regeneration of the riverside, a dark and seedy area that most people chose to avoid. Historically the river has also acted as a physical and social barrier disconnecting south Leeds from the rest of the city. Any plans from the Council were overtaken by the establishment of the Leeds Development Corporation who spearheaded their own riverside regeneration programme by enticing ASDA (now part of Wal-Mart) to build their new headquarters on an abandoned site on the south bank. This low-rise development is evidence of land values that pertained when the offices were constructed in 1988 and it is a measure of the LDC's success that the revalorisation of these once derelict spaces now means that only multi-storey developments can ensure profitability. Gradually regeneration has moved eastwards along the river but, under both the LDC and the City Council, the philosophy of development at all cost has produced a series of rather ad-hoc schemes of little aesthetic quality that have jeopardised the potential for riverside walks and recreation for everyone who lives in the city.

'Leeds', according to architectural critic Kenneth Powell, 'is an architecturally timid city' (2003: 12). The failings of modernism had led to a reaction in the 1980s and in an effort not to repeat the 'mistakes' of the past, a series of unwritten rules produced a plethora of buildings clad in red brick with grey slated hipped roofs – buildings with echoes of the industrial vernacular that were intended to look at home in the city. The so-called 'Leeds Look', a term coined by Powell in an article in which he lambasted the city planners for 'the offence of the inoffensive' (1989: 124-6), was imposed with a dreary monotony, disconnecting form from function and, even where the location cried out for a large, bolder statement as at western end of the Headrow, we have the feeble, lack-lustre Westgate Point (1987). A little further along the Headrow is the multi-coloured confection of the Magistrates Court (1994). Playful it might be (it was locally dubbed 'Legoland'), but unless one wishes to downplay the significance of petty crime, hardly appropriate for a judicial building. Leeds was rapidly becoming something of a laughing stock in architectural circles and early in 1990 a public meeting was called by the Council to discuss 'the Leeds Look'. Attended by 250 people, the architects present used the opportunity to launch a vitriolic attack on the city planners. The days of the 'Leeds Look' were clearly numbered.

Despite Britain being one of the most dynamic architectural centres in the world, what replaced the Leeds Look, were other insipid, gimmicky, developer-led projects. According to Irena Bauman, the Leeds-based architect, the extraordinary boom experienced by the city created 'a licence to build quickly and badly' (2008: 23). When Kenneth Powell came to write his 'New Architecture in Britain', published in 2003, he included only one development in his home city of Leeds, the Cloth Hall Street Apartments, whilst the city's great rival, Manchester merited the inclusion of ten schemes. Only two buildings in Leeds have received RIBA awards: the refurbishment of Brodrick's Corn Exchange and Jeremy Dixon's understated new entrance to the Henry Moore Institute (1993). Despite this lack of architectural plaudits, there have been some worthwhile additions to the cityscape, many of them modest developments such as the Brooker Flynn's frontage for Harvey Nichols (1993) and Carey Jones' elegant in-fill building at 15-16 Park Row. Another Carey Jones' project, Princes Exchange (1999), facing onto the new entrance to the Railway Station, represents an arresting response to an awkward triangular site, and is dramatically lit at night, a development encouraged by the Council's Annual Lighting Awards. The Leeds public enjoy the Sir Basil Spence Partnership's glass-fronted Bourse (1993) for its ever-changing reflections of the 19th century facades along the north side of Boar Lane, and have warmed to No1

City Square (1996-8), perhaps not difficult given the dire, 1960s Norwich Union office block which previously occupied this key site. This new development, also for Norwich Union, was the first major building to eschew 'the Leeds Look': the architects Abbey Hanson Rowe produced a successful compromise between a historicist (in this instance, Art Deco) and functional building in black and white marble, bisected by a glazed elevator shaft under a roof canopy. Striking, Egyptian-styled bronze detailing to the windows enlivens the facades but what really endears No 1 City Square to those who pass by, is the flight of sculpted sea gulls that ascend the front of the building. Whilst other re-inventing cities clamoured after flagship buildings to celebrate their new identity, Leeds appeared more concerned with the money-making aspects of economic regeneration, – indeed the 1999 Vision for Leeds publication even went so far as to reject any need for iconic structures. Five years later the second Vision for Leeds with its focus on 'Going Up A League', and the enhancement of the city's European profile, refers to the creation of landmark buildings, which give expression to a city.

It is hard now to remember that the public reaction against modernism was in part fuelled by a resentment of tall buildings, an antipathy taken on board by planning authorities all over the country. Even in 2000, when Leeds produced its first City-centre Urban Design Strategy, it defined a tall building as any structure above ten storeys. Yet by August 2008, there were eight buildings in the city above twenty storeys and planning approvals for a further three including the landmark buildings of tomorrow, the stunning 52 storey Lumière Building on Whitehall Road, and equally spectacular 'the Kissing Towers' of the Criterion Place development which comprised two interlocking towers of 53 and 33 storeys; both schemes were designed by Ian Simpson of Manchester. Bridgewater Place is currently the tallest building in Yorkshire with 32 storeys. But height is no guarantee of quality as is testified by its 2008 nomination for Building Design Journal's Carbuncle Cup awarded to buildings that are 'so ugly that they freeze the heart' (BBC Leeds 2008). This new turn towards the high rise, quite literally 'Going Up a League' displays an uncharacteristic boldness. At a time when rival cities such as Manchester and Sheffield have been much more nervous about 'a skyline strategy', Leeds' *gung-ho* attitude is largely driven by an economic rationale rather than aesthetic adventurism. It is response to the fact that large development sites in the city-centre have come to an end and to a recognition that the property industry has no alternative but to go upwards or move out of the city-centre. The city architect, John Thorp admitted as much when he justified this policy shift as: 'part of the development of the city and its emergence as a major financial and legal centre' (Rose 2006). A Tall Buildings Design Guide, promised by Thorp in 2006, did not appear until two years later, a sure sign of political disagreements and lobbying somewhere in the system, and, whereas earlier it had been understood by developers that high-rise buildings would be confined to north-south ridge running from the Leeds University's Parkinson Building to Bridgewater Place, the Tall Buildings Design Guide has proposed that other 'gateway locations' would also be appropriate for tall buildings.

Without doubt the most successful and popular aspect of the post-modernist agenda has been the revaluation of the historic fabric of the city, since being a Victorian city is now something to celebrate rather than reject. Historic buildings add variety to the city-centre, bring visitors to Leeds and furnish prestigious premises for certain kinds of businesses. They are the essential ingredients of a heritage industry much celebrated by the city marketers. But in this re-appropriation of history only certain historical periods matter. Although the historian's gaze is shifting to the 1960s and '70s, the city seems only too happy to lose its stock of post-war building. There was an unexpected campaign in 2007 to save the International Swimming Pool (1966), designed by the disgraced John Poulson and the city architect, E.W. Stanley. Once a symbol of civic pride, this archetypal design had certainly been beset with constructional problems, not helped in recent years by lack of maintenance, but it was market forces that finally sealed its fate. Given that we had come to a time when we were all being urged to take more exercise and the fact that the pool occupied an extremely valuable site on the western edge of the city-centre, it was decided to put the land on the market and build a new pool in the south of Leeds – actually paid for largely by Sport Britain so the City Council were winners all round. The losers were those who worked, and increasing lived, in the centre. If not actually demolished, other buildings of the period have been re-clad out of all recognition; for example, nobody is likely to mourn the disappearance of Dudley House (1972) but it is a symbol of the times that what was once the headquarters of the City Housing Department is now a block of luxury flats and offices renamed K2 The Cube (2002).

The chief restoration during this historical turn has been of Victorian/Edwardian buildings: all over the city 19th century mills and warehouses have been converted into apartments, offices, bars and hotels. The city-centre's few remaining churches and chapels are cherished as valuable additions to the cityscape, whereas only thirty years ago they were demolished with little more than a second thought. Some of these conservation projects are modest, such as the refurbishment of some of the yards off Briggate, spaces that go back to 1207, whilst others, for example the Grand Theatre, Brodrick's Leeds Institute (built as a Mechanics Institute and now the city's new museum) and the Corn Exchange also by Brodrick, are magnificent examples of Victorian architecture. Perhaps the highlight of the city's conservation programmes is the stunning Victoria Quarter, now the city's most expensive retailing area. Owned by the Prudential Assurance Company, its costly plans for refurbishment involved the highly contentious roofing over of the former Queen Victoria Street and the projection of that roof structure out onto Briggate with no attempt to integrate the painted steel structure with the adjoining buildings. The developers drove a hard bargain making it clear to the city planners that this was an all-or-nothing project. The result however has been a great success: when work began, only six of original mahogany shop fronts were intact and a good deal of time and money was spent reproducing these, replacing flamboyant metal work and making good damaged faience and marble. The result is that according to Susan Wrathmell in her Pevsner Architectural Guide to Leeds, the County Arcade is '... one of the most beautiful interiors in the city ... glowing with exuberant decoration in marble, mosaic and Burmantofts faience.' (2005:159-61). The doubts of those who objected to the loss of a public street, have been assuaged by Brian Clarke's bright stained glass roof with its abstract patterns of blue, red, green and yellow – no attempt here at Edwardian pastiche – and the cafes with their 'outside/in' ambience which now line the centre of the new arcade. Even the curmudgeonly Alan Bennett is happy to see the County Arcade

> so splendidly restored', though he adds 'if I am honest it's just a bit too smart for me – too done up, and the painted leaves make it look like Christmas all year round – but it's a small price to pay to keep it from the bulldozer. (2005: 522)

Leeds' Post-Modernism: Three Case Studies

(1) Landmark Leeds (1991-2)

The brainchild of Jon Trickett, Landmark Leeds was an ambitious endeavour to change the image of a grim northern city by bringing downtown Barcelona to the streets of Leeds. A group of already pedestrianised Victorian streets to the west of Briggate were selected for this first exercise in post-modern urban design. The purpose of Landmark Leeds was to link the shopping centres to the north and the south with the city's central retailing core by physically breaking up the familiar street lines, creating mini-piazzas for urban spectacle and providing new street furniture, lighting and decorative paving. Planters were to be placed in the wider streets and key junctions were to be marked by sculptures (never executed). A gateway feature was located on the narrow entrance to Briggate, signposting that the pedestrian was about to enter a very different Leeds. Here one might live out the dream of the 24 Hour City: shopping by day and partying at night. Designed by Faulkner Brown of Newcastle, according to Guy Julier,

> The seats, lighting poles and balustrades were doing more than providing a perch or leaning post for weary shoppers, therefore. They seemed to be conspiring to redefine urban identity through their form. (2000: 117-121)

Landmark Leeds also showed how new forms of urban governance dilute democratic inputs into the decision-making process. Even with the Leeds Initiative, the Chamber of Commerce and local shopkeepers on board, the scheme's promoters were nervous that the expenditure of £3.6m on architectural detailing would be regarded as a waste of public money, and would be opposed by local conservation organisations as out of keeping with the character of a Conservation Area largely made up of Victorian buildings. The result was that the redevelopment was kept under wraps. This disturbing saga unfolded on the pages of the Civic Trust Newsletter; in April 1991 readers were informed of 'the erection of a steel pergola along the entire length of Bond Street, Commercial St., Albion St., and Lands Lane. The entrances to the streets are to be marked by 50ft. high gateways in tubular steel and terracotta'. The article goes on to explain that 'Our knowledge of this scheme is very sketchy since the City Council has successfully sidelined our clearly expressed wish to be involved ... the details on the planning application are very limited and virtually incomprehensible'. The author of this chapter can vouch for the latter comment. Sent as a representative of the Victorian Society to the Planning Department for help in interpreting the application, I was informed by a planning officer that the department knew no more about the scheme than I did! In the next edition of the Civic Trust Newsletter, Trust members were told that, at the eleventh hour, the evening before the deadline for comments on the application, the Trust were invited to a presentation by the designers and it turned out that the feature initially thought to be a pergola was a representation of the pattern on the new paving! A year later, in the Daily Telegraph, Kenneth Powell described the scheme as 'out of place' and 'a disgrace to a city full of Victorian craftsmanship', but of course, this was precisely the point of the exercise, to be out of place and to discard the Victorian image of the city.

Having steamrollered the scheme through the planning authorities, Landmark Leeds encountered criticisms from local traders that there was so much 'clutter' in the streets leaving barely enough space for shoppers. In 1995 the 37 aluminium lighting columns were replaced by an under-eaves system and soon the decorative paving, clearly not the work of 'Victorian craftsmen', began to lift, leading to a number of accidents. The Council sued the contractors, the start of fourteen-year litigation. Landmark Leeds was never completed.

(2) Millennium Square (1999-2000)

If Landmark Leeds created controversy, Millennium Square caused outrage in some circles, but, given the Labour Government's policy on social inclusion, the project was presented very differently from the previous scheme. The idea for a new square originated in the Leeds Initiative. It was argued that a square would provide a new focus for the city and create social bonds between residents, as well as providing space for a variety of public events that could be attended by up to 5,000 people. A £5.4m grant was received from the Millennium Lottery Fund and the City Council contributed £6.6 million, some of which it hoped to recoup by selling the revalorised properties it owned in the vicinity of the square. The site proposed for the square was in front of the Civic Hall, a mix of roads, public gardens and car parks which John Thorp, the scheme's designer, had to pull together to create a large open space. Permission was granted by the Highway Authorities to close three of the streets, producing an L-shaped area, but the demands of traffic circulation still meant that, on two sides, the square would face onto busy roads.

Objections to the project were discussed at a series of public meetings at which John Thorp made presentations of the project. The range of criticisms varied:

the idea of underground car parking was successfully knocked back, there were worries about the loss of the public gardens and fears that the square becomes an entertainment centre detracting from the dignity of the Civic Hall. Perhaps more substantially it was argued that the space was unwieldy and lacked any sense of closure; with eight entrances the square would simply leak away into the surrounding streets. Visions were presented of a large, windy, rained-drenched space empty of people: Millennium Square would become a white elephant.

Today most of the critics acknowledge that their fears have not been realised. The dramatic paving of the square has helped unify the space. Seats, trees and planters located on its edges suggest a sense of closure – the trees in their planters can be removed when large events are held in the square, an idea which John Thorp picked up from the Louvre. Another of the conundrums faced by the designer was how to furnish a space that is formal and dignified and yet also responsive to the needs for popular entertainment. The civic identity embodied in the Civic Hall has been emphatically inscribed in this new environment by the placing of large, gilded art-deco styled owls perched on tall obelisks which replicate features of the Civic Hall itself, in positions slightly to the south of the hall's portico. At the southern edges of the fan-shaped 'arena area' is the Mandela Garden which includes a water feature (to help soften the sound of traffic) and the prize-winning garden from the 2006 Chelsea Flower Show. Within this garden area is a 16 ft. sculpture, 'Both Arms', representing the spirit of reconciliation and created by the Leeds-born Kenneth Armitage. The facilities which service the entertainment functions of the square are located underground, except for a tower at the north eastern corner which has been not very successfully incorporated into a grenade-like sculptural structure, 'Off Kilter', designed by the artist Richard Wilson and financed by the Henry Moore Foundation. The tower is attached to a terrace bar, a perfect vantage point from which to observe the to-ings and fro-ings in the square itself.

Millennium Square is now an accepted part of the city's public life, it is location for a variety of public celebrations and events: the switching on of the Christmas Lights, the Christkindel Markt, firework displays on New Year's Eve, a skating rink in January and February, the St Patrick's Day Parade and in addition to a number of concerts scattered throughout the year, the BBC's 'Big Screen' regularly transmits a range of programmes from ballet and operatic performances to key sporting events. If ever Leeds United were to win a major football prize, the celebrations would take place in this square. The recent opening of the new City Museum on the eastern side of Millennium Square can only serve to bring more people into the space and add to its popularity.

As anticipated by the Leeds Initiative and the City Council, Millennium Square did become a catalyst for the regeneration of the surrounding area and £120m has been spent by the private sector on perimeter sites. On the southern edge of the square, the Electric Press Building of 1866 houses a large bar and Leeds Metropolitan University's School of Film and Media, the West Riding Carriage Works (1848) with its courtyard now glazed over, is home to a variety of restaurants and bars and adjoining these historic buildings is the new-build Carriage Works Theatre supported by circular columns with more cafes and bars below, their tables and chairs spilling out into the square.

(3) Clarence Dock (2002-8)

As in cities like Manchester and Bristol, the regeneration vision was to create a new residential and office area around a major tourist attraction, in this instance, the Royal Armouries Museum. Little else was achieved before the LDC's demise in 1995 but the challenge of developing this area was taken up by the Leeds Initiative. By the turn of the century, at the height of Leeds' economic boom, waterfront locations in the more immediate city-centre were disappearing with the result that investors' attention began to shift eastwards. The New Dock (1840-3) just to the west of the Royal Armouries lay virtually derelict. A master plan for the area by Browne Smith Baker would result in a more coherent development here than elsewhere along the river. The plan focused on the dock itself, a rectangular basin 100 metres by 50 metres with bold stone quay walls. Locally this had been known as the Potato Dock and a truly post-modernist gesture would have been to retain this nomenclature but, with a £260m development at stake, the developer, Crosby Homes, played safe by choosing the new name of Clarence Dock. Designed by the architects, Carey Jones, the scheme includes 1124 luxury apartments (none designated as 'affordable housing') plus a limited number of berths for houseboats, bars and restaurants, shops and offices and a casino. The latter opened in 2008 costing £13.5m, and comprises a main gaming room with a capacity for a thousand punters, above is a poker room and small cinema, and a number of bars and two restaurants.

Although not distinguished in architectural terms, the space is dramatic quite unlike any other in Leeds. You could be in Rotterdam or, for that matter, Barcelona. All of the apartments were sold before their completion but how far they are occupied is unknown. Certainly at the time of writing, the main office block remains largely untenanted. A major problem, apart from its setting in an urban wasteland, is the site's inaccessibility from the city-centre. The casino organises its own bus service at the weekends and although for a time Leeds City Cruisers provided a waterbus service, this has now ended because of lack of customers.

In two television programmes, in 2006 and 2009, Maxwell Hutchinson, a former president of RIBA, has lambasted Clarence Dock, describing it as an urban desert, soulless and depressing:

> At Clarence Dock, they don't go in for schools or churches ... but there's a shiny new casino. There is only one small supermarket, but if you want to eat out there is plenty of choice. In two or three decades these shiny new buildings will be following Quarry Hill down the spiral of decay. (BBC Insideout 2009)

Despite a flashy formal launch in the autumn of 2008 by Gok Wan, the fashion designer, in the short term the prospects for Clarence Dock appear grim but maybe in this period of economic depression, the casino will prove to be the saviour of the reinvention of this part of the city. Who knows?

The Two Speed City and other Failures

A flourishing financial services sector and glitzy buildings are not everything. The city-centre is still surrounded by areas of poverty, though this concept is subsumed under the less disturbing notion of 'social exclusion'. The 'trickle-down' model of city-centre revitalisation remains a chimera for the '150,000 or 10% who live in areas that are officially registered as being amongst the most deprived in the country' (Vision for Leeds 2004)). Some commentators doubt whether, irrespective of 'the narrowing the gap' vision, this objective is compatible with the city's ambition 'to go up a league'. For Chatterton and Hodkinson, for example, 'it is simply impossible for Leeds 'to narrow the gap' as the city-centre continues to become the exclusive playground for tourists, students, the wealthy and the professional business class (2007: 25). In its strongest form the concept of a Two Speed City describes a model where division between 'the haves' and 'have-nots' is not merely contingent: the rise of the new service class in well-paid posts requires large numbers of people who have few alternatives except to work in low skilled and low-paid clerical and service jobs. Rather than 'narrowing the gap', there is evidence that social polarisation in Leeds is widening: the Economic Handbook for Leeds 2008 records that although unemployment levels declined between 1997-2005, during the following two years they increased above the levels for the region and the country as a whole. Correspondingly in the twelve months leading up to January 2009, claimant levels rose by a staggering 54%. One in three children in the inner city are growing up in households dependent state on benefits and information placed in the Library of the House of Commons (December 2008) revealed that Hilary Benn's Central Leeds constituency has the 16th highest proportion of children in this category out of the 628 constituencies in the country. This is not the place to recount endless statistics of deprivation. For a graphic picture of life on the social margins, read Bernard Hare's 'Urban Grimshaw and the Shed Crew' (2005) which in stomach-churning, Dickensian detail presents an alternative view of Leeds.

Not all is well in the city-centre either. Wayne Hemingway, the founder of 'Red or Dead', ruffled feathers at the council-led Leeds City Centre Vision conference in 2008 when he suggested that 'the buildings going up today are totally indistinguishable from each other and will have to be pulled down in less than fifty years. He went on to say that Leeds did not offer, 'cradle to grave living, and that although loads and loads of good words have been written about the future, the city was failing to deliver and could not live up to its branding Live It, Love It' (Waite 2008).

Some parts of the city-centre have proved resistant to programmes of civic beautification. The Neville St Underpass, a gloomy cavern used by 21,000 pedestrians every day, is also the main route into the city from the M1 and M62 motorways. Kirkgate, owned by EMCO, is one of the oldest streets in the city yet it remains one of the most unsavoury – a street of run-down Georgian and Victorian buildings (including the White Cloth Hall of 1711); its crumbling shops do, however, serve an important social need. Most of the shops are still open, occupied by retailers and café-owners who would never be able to afford the high rents charged in other parts of the city-centre. Even more dreary, despite its location off Briggate, is the 1960s New Market Street Arcade owned by a Liechtenstein property company. The mega-developments which have been encouraged by Leeds policy-makers have meant that small spaces in the centre have been neglected despite the fact that that such patterns underpin the vitality of the European cities that Leeds wishes to emulate. Many of the yards off Briggate and the Headrow remain unkempt and uncared for, including Lambert's Yard with its 16th century timber framed, jettied house, the oldest surviving building in the city. In Bramley's Yard of the Headrow, we find Big Lil's, a bar whose clientele live in a different world from those who frequent the stylish bars found in other parts of the centre.

The problematic experience of the City Living phenomenon has already been outlined. The theme of 'slums of tomorrow' is now married to anxieties about the effects that the economic downturn will have on property values, maintenance levels and occupancy rates. Despite the 'Knightsbridge of the North' tag, Leeds has more 'cheap and cheerful' retail outlets than up-market stores. In this context it is notable that the latest retail premises to come on the market, Broad Gate, which replaced the old Lewis's department store, is occupied by a branch of Sainsbury's, a large TK Maxx Discount Designer Shop and an Argos Store. Jon Trickett's dream of a 24 Hour European city has backfired. According to Pete Connolly, small property developer and restaurant and bar owner in the city:

> We are now teetering on the edge of a 24 hour booze economy. The dream of
> a continental-styled cafes and entertainment for everyone has proved just that
> – a dream. It's booze everywhere. We've been hi-jacked by mega pub chains.
> (Wainwright 2005)

The night-time economy, once seen as making the city-centre safer, has had precisely the opposite effect. A drunken hedonism has invaded the city and its effect is to deter older people, ethnic minorities and the disabled from visiting the city-centre. Despite additional policing and the presence of 270 CCTV cameras (there were just 19 in 1996), alcohol-related incidents cost the city about £275m a year, costs to the NHS are more than £23m whilst the criminal justice system spends almost £160m dealing with vandalism, drunkenness and violence (Chatterton 2007: 146).

An over-dependence on what Americans call the Fire Sector (finance, insurance and property development) will mean that Leeds will suffer more than some other places in the current credit crisis. Unfortunately for its citizens there was never a

Plan B. For months, the local press has carried an incessant stream of stories about lay-offs and plummeting house prices. Of the 12,770 apartments in the pipeline at the beginning of 2008, 7,000 have already been scrapped, the take-up of office space is down by 23 percent and in this new age of austerity, despite being urged to spend more to save the global economy, the retail sector in January 2009 reported its fourth consecutive quarter of decline. Despite reassuring speeches by council leaders and the Chamber of Commerce of 'temporary blips', 'streamlining' and 'mixed pictures', the development juggernaut has been halted. Many of the city's high-profile projects have been cancelled including the Kissing Towers and the Lumière Building – work stopped here abruptly in July 2008 leaving a rather spectacularly derelict space, with giant piles driven into the foundations from which it had been expected that gleaming glass towers would rise up. The City Council has announced that the Spiracle building, a 24-storey apartment block planned to replace the International Pool, will temporarily become a car park (Waite 2008). As mentioned above, the future of the Eastgate/Harewood Shopping Centre must be in doubt. Emblematic of this unravelling of Renaissance Leeds was the shock news on 26th February 2009 that K.W. Linfoot PLC, which spearheaded Leeds city living and was the joint developer of the Lumiere had gone into liquidation (Yorkshire Evening Post 2009).

It may be a straw in the wind but the surprise decision by the local authority's Executive Board to build and own the planned 12,500 seat City Arena suggests that the days of the public/private partnership may be coming to an end. It had been anticipated that either Montpellier Estates or the GMI Group would be selected to develop the Arena on sites in the south of Leeds but one of the reasons given for Council's change of plans was that the cost of public subsidy required by the private sector had rocketed. The project will now be financed by council capital receipts, borrowing on the back of rental income received from the Arena operator, SGM, and grants from Yorkshire Forward. The Council decision also involves a change of location; the Arena will now be built on the northern edge of the city-centre close to the Merrion Centre. The five-acre site is owned by the City Council and Leeds Metropolitan University; Leeds Met had been hoping to sell their land for a £275m housing scheme but in August 2008 the developer pulled out. The decline in the demand for housing and falling land values, plus the University's own complex financial position, has put the Local Authority in a position where it can afford to buy Leeds Met's share of the site (Yorkshire Evening Post 2008). It is perhaps too early to talk of 'the Return of the State' in the local context but the retreat from the market and the revival of the state nationally demonstrates that an extraordinary political shift is taking place.

Conclusion: the Balance Sheet

Over the centuries cities and towns have had continuously to reinvent themselves in order to survive: in Leeds historians record the changes from a medium-sized commercial and retailing centre in the eighteenth century to a centre for industrial manufacturing in the Victorian period. Understanding the history of the twentieth century Leeds is far more complex. Here we see a number of distinct periods: until around 1914, it was 'business' as usual and in the inter-war period, there were aspirations for modernity which largely focused on making Leeds a more egalitarian place. The struggle for modernity took a somewhat different turn in the years after the Second World War when the erosion of the city's manufacturing base and the related modifications in the social structure produced some of the changes which some now wish to claim for post-modernity. Leeds was searching for a new post-industrial identity and, from the early 1960s, councillors and their officers began to court developers. Geoffrey Thirlwell, the city's Chief Engineer from 1964, was an enthusiast for the comprehensive redevelopment of the city-centre and for linking Leeds to the motorway network, believing that it would bring investment into the city. There was an explosion of office-building long before the City Council gloated about being the financial and business services centre of the north of England. The long post-war boom gave rise to the first shopping centres, and the aim of pedestrianisation was to attract consumers into the city-centre. For Alan Bennett this was the period when:

> ... avarice and stupidity got the wheel of the bulldozer. They called it enterprise and still do, but the real enterprise would have been if someone in 1960 had had the clout and imagination to say 'Let us leave this city much as it is, convert it perhaps, replumb it, but nothing else'. (2005: 501)

Men like Thirlwell and Frank Marshall, Conservative Leader of the Council between 1967 and1972, were as obsessed with 'place-making' as more recent politicians; then as now there was a determination to discard the image of a grim northern Victorian city and promote the vision of a dynamic modern city. It was in the 1980s that despair and demoralisation set in: the rise of neo-liberalism, the discrediting of the state, both national and local, changes embodied in the Thatcher governments, exercised a hegemonic influence that produced today's born-again city. Policy-makers seem to have bought into a popular image of the North of England, what the playwright Dennis Potter called 'eh-bah-goom heritage' (Russell 2004: 2), a narrative that has grown up over the last two hundred years. At times Leeds policy makers exploited this negative representation of grime, decay and philistinism to prise money from central government but the myth also invaded their own consciousness, leaving them bereft of the confidence and the will to temper the demands of developers and to pursue strategies which were embedded in a distinctive sense of place.

On many fronts the new post-modern Leeds represents a success story: a remarkable economic revival, the expansion of cultural facilities, the regeneration of the waterfront and refurbishment of historic buildings. Any audit, however, of Renaissance Leeds must necessarily be subjective. The city of the mind cannot be contained by tables of statistics demonstrating prosperity and levels of investment in the city-centre. The city's remarkable economic turnaround has certainly benefited many but those living in the inner-city and on periphery council estates, victims of the bonanza, look in bafflement at a city they no longer recognise, nor feel that they belong to. Whilst most would agree that the aesthetic appearance of the city-centre had improved, others find the city too sanitised, over-choreographed in ways that are destructive of individual character and atmosphere. New shopping complexes and apartment blocks have often led to the privatisation of what was once public space – even Millennium Square, 'the people's square', is inaccessible to the general public on ticket-only occasions. Michael Paraskos in his pamphlet, 'Leeds: Barnsley of the North' has poked fun at the city's claims to being the Knightsbridge of the North or, even less appropriately, the Barcelona of the North. What's wrong with these claims?

> is not their breath-taking, bombastic absurdity; it's the small-minded lack of confidence which can only express Leeds' worth as a city by comparing it to places it's clearly nothing like. (Oliver Cross, 2006)

Such inflated rhetoric may well cosset the egos of policy-makers and developers, but because it confuses rhetoric with reality, it gives expression to an unattainable goal, no amount of mediocre new building and street furniture can turn a city into something it can never be.

Local historians tell us that Leeds has a history of weathering economic depressions – an argument eagerly seized upon by today's politicians – but in the face of a local economy now based on financial services and consumption, the present economic meltdown may sweep away the city that many imagined Leeds to be. When real life changes, it also leads to changes in academic paradigms. Notions of post-modernity may well belong to a particular historical period that is now at an end but what will replace it remains to be determined.

References

Bauman, I. (2008). Opinion, *Architects' Journal*, 227(16), 23.

BaumanLyons. (2009). *How to be a Happy Architect*. London: Black Dog Publishing.

BBC Insideout Yorkshire and Lincolnshire. (2009). Slums of the Future? [Online] Available at: http://www.bbc.co.uk/insideout/content/articles/2009/01/28/yorks_lincs_s15_w3_ [accessed 11 March 2009].

BBC Leeds (2008). A Bridgewater too Far? [Online] Available at: http://www.bbc.co.uk/leeds/content/articles/2008/10/07/places_carbuncle_cup_feature.sht [accessed: 11 March 2009].

Bennett, A. (2005). *Untold Stories*. London: Faber and Faber Ltd.

Burt, S. and Grady, K. (1994). *An Illustrated History of Leeds*. Derby: Breedon Books.

Chatterton, P. (2007). Life in the Frontier Zone, in *Around Leeds: A City Centre Reinvented*, by J. Stillwell and R. Unsworth, Leeds: University of Leeds Press, 145-6.

Chatterton, P. and Hodkinson, S. (2007). Leeds: Skyscraper City. *Yorkshire and Humber Review*, Spring, 30-32.

Chatterton, P. and Hodkinson S. (2007). Leeds: an Affordable, Viable, Sustainable, Democratic City? *Yorkshire and Humber Regional Review*, Summer, 24-6.

Chatterton, P. and Hodkinson, S. Leeds Regeneration. Autonomous Geographies [Online] Available at: http://www.autonomousgeographies.org/leeds regeneration [accessed] 11 March 2009.

Chatterton, P. and Unsworth, R. (2004). Making Space for Cultures(s) in Boomtown. Some Alternative Futures for Development, Ownership and Participation in Leeds City Centre. *Local Economy*, 19(4), 361-79.

Crabtree, P. (2008). Leeds' Planning and Development Ambitions, in *Around Leeds: A City Reinvented*, by J. Stillwell and R. Unsworth. Leeds: University of Leeds Press, 21-2.

Crosland, A. (1956). *The Future of Socialism*, London: Jonathan Cape Ltd.

Cross, O. (2006). Nothing to Brag About. *Yorkshire Evening Post* 5 May [Online] Available at: http://www.yorkshireeveningpost.co.uk/features/Nothing-tobrag about.1485509.jp [accessed 31 March 2009].

Dutton, P. (2003). Leeds Calling: the Influence of London on the Gentrification of Regional Cities. *Urban Studies*, 40(2), 2557-72.

Gunn, S. (2000). Ritual and Civic Culture in the English Industrial City c.1840-1914, in *Urban Governance Since 1750*, edited by R.J. Morris and R.H. Trainor. London: Ashgate, 226-41.

Hardcup, T. (2000). Re-imaging a Post-industrial City: the Leeds St Valentine's Fair as a Civic Spectacle. *City*, 4(2), 216-31.

Hare, B. (2005). *Urban Grimshaw and the Shed Crew*. London: Sceptre.

Hartley, O. (1980). The Second World War and After, in *A History of Modern Leeds*, edited by D. Fraser. Manchester: Manchester University Press, 437-61.

Haughton, G. and Williams, C.C. (1996). *Corporate City? Partnership, Participation and Partition in Urban Development in Leed*s. Aldershot: Avebury.

Haughton, G. and While, A. (1999). From Corporate City to Citizens City: Urban Leadership After Local Entrepreneurialism in the United Kingdom. *Urban Affairs Review*, 35 (3). 3-23.

Julier, G. (2000), *The Culture of Design*, London: Sage.

Leeds City Council. (1926). *Leeds Tercentenary Handbook.*

Leeds City Council. (1971). *Project Leeds: Leeds Motorway City of the Seventies.*

Leeds City Council. (2000). *Leeds – City Centre Urban Design Strategy.*

Leeds City Council. (2008). *Tall Buildings Design Guide.*

Leeds City Council. (2008). *Leeds Economic Handbook 2008.*

Leeds Civic Trust. (1972-2009). *Newsletters.*

Leeds Culture. (2002). *The Cultural Strategy for Leeds.*

Leeds Guide. (1806). *Leeds: Edward Baines.*

Leeds Initiative. (1999). *Vision for Leeds.*

Leeds Initiative. (2004). *Vision for Leeds 2004-2020.*

Leeds Surrealist Group. (2004). City Hung, Drawn and Quartered. *Manticore,* Summer/Autumn.

McClea, N. (2008). Improve the Image As Well As the Physical and Social Fabric of The City, in *Around Leeds: A City Centre Reinvented,* by J. Stillwell and R. Unsworth, Leeds: Leeds University Press.

Morris, R.J. (2007). Towards a History of the Twentieth Century. Paper to the Thoresby Society: Leeds, 7th February 2007.

Morris, R.J. (2008). Whose Time and Whose Place: Searching for the History for the History of Twentieth-Century Leeds. *Publications of the Thoresby Society, Second series,* Vol.18 (Miscellany), 1-17.

Old Leeds Cropper (1868). *Old Leeds: Its Byegones and Celebrities.* Leeds: H.W. Walker.

Paraskos, M. (2004). *Leeds: Barnsley of the North.* Leeds: New Leeds Arts Club.

Pevsner, N. (1959). *The Buildings of England: Yorkshire: West Riding.* Harmondsworth: Penguin Books.

Powell, K. (1989). The Offence of the Inoffensive. *The Architect's Journal,* 3rd May.

Powell, K. (1992). What Have They Done to Our Cities? *The Daily Telegraph,* 11th April.

Powell, K. (2003). *New Architecture in Britain.* London, New York: Merrell.

Powell, K. (2005). White Rose Blooms. *Building Design,* 4th November 2005, 12-15.

Ravetz, A. (1974). *Model Estate: Planned housing at Quarry Hill, Leeds.* London: Croom Helm.

Rimmer, W.G. (1967). Occupations in Leeds 1841-1951. *Publications of the Thoresby Society,* L, 158-78.

Rose, J. (2006). Leeds takes the high-rise lead. *Building Design* [Online] 29 September. Available at: http://www.bdonline.co.uk/story.asp?storyCode=3074334 (accessed 11 March 2009].

Russell, D. (2004). *Looking North: Northern England and the National Imagination.* Manchester: Manchester University Press.

Sawyer, M. (1993), The Economy of Leeds in the 1990s, in *Leeds City Business,* edited by J. Chartres and K. Honeyman. Leeds: Leeds Univeritiy Press, 270-80.

Spink, J. and Bramham, P.(1999). The Myth of the 24-hour City, in *Policy and Publics: Leisure, Culture and Commerce,* edited by P. Bramham and W. Murphy. London: Leisure Studies Association 65, 139-51.

Stillwell, J. and Unsworth, R. (2004). Creating Prosperity: Jobs, Businesses and Economic Initiatives, in *Twentieth Century Leeds: Geographies of a Regional City,* edited by R. Unsworth and J. Stillwell, Leeds: the University of Leeds, 167-90.

Stillwell, J. and Unsworth, R. (2008). *Around Leeds: A City Centre Re-Invented. Leeds:* Leeds University Press.

Strange, I. (1996). Pragmatism, Opportunity and Entertainment: the Arts, Culture and Urban Economic Regeneration in *Leeds, in Corporate City?*, edited by G. Haughton and C.C. Williams, Aldershot: Avebury, 135-52.

Tickell, A. (1996). *Taking the Initiative: the Leeds Financial Centre, in Corporate* City, edited by G. Haughton and C.C. Williams. Aldershot: Avebury 103-118.

Thompson, B. (1971). *Portrait of Leeds*. London: Robert Hale.

Trickett, J. The 24 Hour City. *Regenerating Cities*, 6, 9-11.

Unsworth, R. (2005). *City Living*. Leeds: Kevin Linfoot PLC and The University of Leeds.

Unsworth, R. and Nathan, M. (2006). Beyond City Living: Remaking the Inner Suburbs. *Built Environment*, 32(3), 235-49.

Unsworth, R. and Stillwell, J, (2004).*Twenty-first Century Leeds: Geographies of a Regional City.* Leeds: The University of Leeds Press.

Wainwright, M. (2005). The Dream of Continental Cafes for Everyone has Proved Just that – a dream. *The Guardian* 22nd January.

Wainwright, M. (2009). *Leeds: Shaping the City.* London: RIBA Publishing.

Waite, R (2008). Hemingway Attacks 'unsustainable' Leeds. *The Architects' Journal* [Online] 1 February. Available at: http://www.architectsjournal.co.uk/ news/dailynews/2008/02/hemingway_attack_leeds.html (accessed: 13 January 2009).

Waite, R. (2008). Leeds Tower Axed in Housing Market Fall. *The Architects' Journal* [Online] 24th April. Available at: http://www.architectsjournal.co.uk/ news/dailynews/2008/04/leeds_tower_axed [accessed: 13 January 2009].

Ward, K. (2003). The Limits to Contemporary Urban Redevelopment. 'Doing' Entrepreneurial Urbanism in Birmingham, Leeds and Manchester. *City*, 7 (2), 199-211.

Waterhouse, K. (1994). *City Life: a Street Life*. London: Hodder and Stoughton.

Wrathmell, S. (2005). *Pevsner Architectural Guides: Leeds*. New Haven, London: Yale University Press.

*Yorkshire Evening Post. (*2008). Shock Leeds Arena Decision.[Online] Available at: http://www.yorkshireeveningpost.co.uk/news/Shock-Leeds-Arena-decisio n.4661128.jp [accessed:11 March 2009].

Yorkshire Evening Post. 26th February (2009). 1, 8-9.

Chapter 4

Mission or Pragmatism? Cultural Policy in Leeds Since 2000

Jonathan Long and Ian Strange

There's no doubt that in the last ten years that the importance of culture has gone way, way up the agenda. [policy partner]

Leeds' cultural offering is quite extraordinary, and being invested in, in quite an extraordinary way. [regional officer]

Introduction

If you were to ask cultural commentators which cities in the UK have used cultural policy to lead development, they might choose: Liverpool, European City of Culture, 2008; Glasgow, European Capital of Culture, 1990; London under the GLC; Manchester; Birmingham; Newcastle, forgetting that much of the development has been across the river in Gateshead; Sheffield; or even Huddersfield/Kirklees – but not Leeds.

It became fashionable to dislike (even revile) Leeds in the 1970s, typified through a popular hatred of Leeds United football club in the Don Revie era, and perhaps some of that rubbed off on outside perceptions of the city. However there are those who appreciate Leeds. For example, in his recent search for 'the North', the broadcaster Stuart Maconie writes enthusiastically about a city that is abuzz: 'Britain's most improved city, the gentrified jewel of the new north, the Barcelona of the West Riding' (Maconie 2007: 202). Despite a childhood scarred by memories of television programmes from Leeds City Varieties he finds the city handsome and sexy. He extols the virtues of his hotel, the fusion restaurants, the curved galleries of the Corn Exchange, the fabulous Victorian town hall, the superb arcades (including the 'singular and stylish' additions), and can not understand why Alan Bennett doesn't eulogise more about Leeds Market. He notes that the city art gallery has the finest collection of 20th century art outside London and revels in how the Kaiser Chiefs have managed to incorporate Leodensian into one of their lyrics. 'Of all the great northern civic rebirths of the last 20 years Leeds has been the most talked about because it's been the most unexpected' (Maconie 2007: 204). This is some homage from a Lancastrian.

So, how do we explain on one hand that Leeds fails to register on the list of those cities using culture to regenerate themselves, while on the other some popular cultural critics have lauded a cultural renaissance in the city? Part of the explanation might be that writers, including academic researchers, (in Leeds and beyond) have given it scant attention. Part might also be because some elements of the city have yet to reconcile themselves with their incorporation into Leeds (through local government reorganisation in 1974), so there are ready detractors from within. Ian Rankin likes talking about the two sides of Edinburgh, but, as we might say: "Ey lad, is that t' best tha can do? We've got twa sides ten times ower." Leeds has long been diverse, not dominated by any single industry, be it steel or textiles or shipbuilding, which in recent decades has allowed it to ride out the worst effects of economic downturns. Its population is diverse also, having received successive rounds of immigration, but no single group has been on the scale (proportionately) that some cities have seen.

Leeds' 'cultural offer' is similarly varied. There is little doubt that any audit of cultural activity in the city would look impressive, but then Leeds is the second biggest local authority in the UK[1] so it ought to. This 'offer' covers all aspects of culture, and is contributed to by many different organisations, some of which rarely enter deliberations over cultural policy, such as the Leeds Festival, which its organisers (not from Leeds) boast is 'still the North of England's only [rock] music festival'.

In the UK, cultural policy at sub-national level has been led by local authorities, yet over the past 20 years or so local authorities have had their powers curtailed by successive governments at Westminster. Of course, cultural life exists beyond the ambit of formal cultural policy-making, though much of it is still sanctioned through local authority regulation. For example, the Carling Music Festival is now firmly established in the summer festival diary and the pub music scene has revived despite tighter licensing regulations (perhaps as a response to the downturn in drinks sales).

In thinking about cultural policy in Leeds, the challenge we set ourselves was to investigate key drivers of cultural policy and cultural policy change. We were less concerned with patterns of provision than with why and how policy came to be, how things have changed and how Leeds has come to be positioned in relation to other places. We had previously been involved in separate related studies of Leeds in the early and mid 1990s (Bramham *et al.*, 1994, Strange 1996). This time, to supplement all the written material we had available, we conducted a programme of interviews with 18 people we knew to be influential in policy terms or who were recommended to us by earlier respondents. These interviews lasted about an hour, though some were nearer two, and were supported by less formal discussions with others involved in Leeds culture.[2] Some respondents were initially

---

1 In population terms only Birmingham is larger.
2 Quotes from our respondents appear in italics to distinguish them from quotes from the literature.

nervous that they were going to be tested on their knowledge of the contents of the cultural strategy. Our concern though was to learn from their experience and standpoints. This produced a predictable diversity of views, though often framed within a similar narrative about the nature of cultural provision in the city, suggestive of a neo-liberal commitment to using culture as a key ingredient of city revitalisation.

Major Policy Strands

Policy for the city is set within the context of the *Vision for Leeds*, devised through a fashionable consultation process and first published in 1999; the current (second) version covers the period 2004-2020 (Leeds Initiative 2004). It characterises the challenge for Leeds as:

- Going up a league – becoming an internationally competitive city, 'the best place in the country to live, work and learn, with a high quality of life for everyone'
- Narrowing the gap to improve the position of disadvantaged people and communities
- Developing Leeds' role as the regional capital

Within the context of culture this is expected to entail providing cultural opportunities for everyone, developing talent, achieving recognition as a leading artistic, cultural and sporting city, and creating first class cultural facilities.

Cultural Strategy: Still Opportunistic and Pragmatic?

Informed by our interviews we contend that the hallmarks of Leeds' cultural strategy might reasonably be taken to be opportunism and financial pragmatism. Today, cultural policy in the city still bears the imprint of the Labour administration that held control from 1980 to 2004. Throughout this period, the general approach to cultural policy and provision can be interpreted as delivery-focussed and patriarchal in the tradition of welfarism and Labour patronage. Certainly, the significance afforded to culture and the emphasis of policy have fluctuated over time, but these are enduring themes to which we shall return. Two previous attempts to produce a cultural strategy for the city had failed to make it through to full council approval. But the insistence of the Department for Culture Media and Sport (DCMS 2000) that all councils should produce a cultural strategy finally overcame this reticence with a written strategy produced for the period 2002-7. Leeds' Cultural Strategy 'seeks to create an environment that enables people to realise their full potential and to feel the positive and creative life of the city is reflected in their everyday life'. Further, it will 'promote the cultural wellbeing of the area, be inclusive and reflect the way of life of communities, taking into account the geographical identity, local history and character of the area' (Leeds Cultural Partnership 2002: 10-12). The strategy is founded on a set of core values:

- fun and enjoyment
- belonging
- celebrating diversity
- realising aspirations
- working in partnership
- promoting equality of access.

The Cultural Strategy comments that 'culture does not belong to large institutions' (Leeds Cultural Partnership 2002: 12). However, throughout our research, when asked about the key elements in Leeds' cultural policy, respondents typically answered in terms of the elite, high-profile provision of major institutions like Opera North, West Yorkshire Playhouse, Northern Ballet Theatre, the Royal Armouries, the art gallery and the new city museum. In light of the city council's substantial financial commitment, this is perhaps not surprising, but it is also the kind of provision that people have learned is necessary for a major European regional city, attracting employers and business en route to becoming 'an internationally competitive city'. Although not necessarily at odds with, it is certainly in marked contrast to, the expansive, inclusive tone of the published strategy document intended to address the agenda of 'narrowing the gap'. The visual imagery of the document is not difficult to read: there are two images of public sculpture, an exhibition, a representation of creative industries and a dance performance; these are supported by a street performance and an amateur dramatic production as well as public involvement in a professional production (Carnival Messiah), people attending Opera in the Park, watching Leeds Rhinos (rugby league) and a cricket test; but then there are 21 images of 'ordinary citizens' engaged in leisure. As one of the senior officers suggested, this is a policy document designed to convey the sense of layering of cultural activity and provision. At an operational level this may be so, but the balance of attention within the policy discourse is on the larger-scale provision, or what has previously been described as searching for big hits (Strange 1996).

To avoid being misunderstood, we should explain that when our respondents were questioned more directly there was no hesitation in discussing the wider social remit of the city's cultural agenda; but the more salient elements of their comments tended to refer to 'prestigious' institutions or projects. However, the point was made that the idea of 'going up a league' does not have to be at odds with 'narrowing the gap':

> *I've never tried to think we're either dealing with excellence or we're dealing with access. Take Opera North as a good example. We give Opera North nearly a million pounds a year, but part of that is a service agreement not just for*

productions on the stage of Leeds Grand Theatre; it's to go out and work in communities in Leeds, in some of the most challenging environments you can imagine to generate an interest in the arts. [senior officer]

The potential contradiction at the heart of a bilateral approach to cultural policy designed to foster both economic competitiveness and social justice was clear for some of those we talked to. One councillor expressed concern that sending well-intentioned arts groups into deprived communities meant taking them some culture '*like a dollop of ice cream*', echoing the arts in the community versus community arts disputes. Another respondent was concerned about the confusion of priorities:

I'm a great believer in what Confucius said: 'Don't chase two rabbits because you won't have any tea tonight'. I paraphrase, but I think it's important to decide which one you're going for. You can't do both successfully. The current sporting equivalent is increasing participation and promoting excellence. The two things are connected, undeniably, but you can't do either properly without focusing on one of them. To go full out to be excellent, inevitably your eye gets taken off the wider community engagement. Giving them both equal status isn't going to work. It's not that you exclude other things, but when it comes to a choice you go with that one, and the next time it comes to a choice you make the same decision. I'm not saying totally exclude it, but if it comes down to being able to do only one thing today then go for excellence and if that means shutting down services ... well the money could be better invested ... You can't go for the community thing and at the same time go for being a world class northern city. [regional officer]

The financial pragmatism at the core of the current approach to policy is just as clear as the focus on big hits and infrastructure. Two major projects initiated early in the new millennium were given as examples of that financial pragmatism. Once the Civic Theatre (formerly the Mechanics Institute) had been identified as in need of overhaul, alternative uses were considered for the building. The outcome was that the council took money known to be available from the Heritage Lottery Fund to provide Leeds with a museum. Although this may not have been the best solution regarding the provision of a (much needed) museum, it did have the virtue of retaining and repairing a significant component of Leeds' heritage in the shape of a building designed by Cuthbert Brodrick, who was also the architect of the Town Hall[3] and the Corn Exchange. The second example is the replacement for the international pool. There were many who wanted it to be retained on its city-centre site, but demolition and rebuild would have left swimming and diving clubs without a facility in the interim. Instead, Lottery funding was used to build

---

3   The final cost of the Town Hall was three times the original estimate. Some things don't change.

a new aquatics centre at the John Charles Centre for Sport on the south side of the city. The funding proposal itself estimated that this solution would lead to 147,000 fewer swims per year, but, on the other hand, it allowed capital assets to be realised on the city centre site of the old pool.

Our interviewees suggested that the city's decision in favour of a new arena and upgrading the town hall rather than building a state-of-the-art concert hall was a populist response. More likely it is a continuation of Leeds' financial pragmatism, to which the preceding two examples are testament. The financial climate is such that a new development on this scale cannot be constructed with public funds alone; a major contribution at least is also needed from private finance. Hence the need for a commercial return and that is more likely to be realised from an arena with a capacity of 12,000 than a concert hall with a capacity of 2,000 and the product is more readily available. It was also felt by respondents to be a 'must-have' in staking Leeds' claim to rank alongside other major cities.

Financial pragmatism comes close to substituting for strategy:

> *A lot of the discussion I see is effectively* [about] *how to maintain what we've got already. The cultural programme is effectively looking each year and saying "Well how do we support what we're already doing? How do we maintain that? Are there new things that are coming along?" and then beyond that there is a significant amount of fire-fighting.* [policy partner]

One senior politician observed: *The cultural strategy for Leeds is currently not a strategy at all; it's a series of capital schemes.* Partly because of the financial environment, Leeds does not have the 'signature' new buildings designed by internationally-renowned architects that attract headlines and awards. Instead, using a large part of the cultural budget to refurbish the Grand Theatre, Town Hall and City Varieties, and turn the Mechanics Institute into a city museum manages to conserve significant elements of the city's built, Victorian heritage. This in itself might be seen to offer a distinctive flavour to Leeds' cultural infrastructure, but it is an approach not without its critics who question whether, if Leeds is supposed to be 'going up a league', this can be achieved with the refurbishment and conservation of outmoded 19th century Victorian buildings.

More generally, some respondents complained of a lack of cultural imagination. One episode is illustrative here. Before the Angel of the North (created by Anthony Gormley and erected in Gateshead) was considered, there was a proposal in the mid-1980s for Leeds to construct a signature piece of public art that would greet those arriving by rail from the north, west and south; dubbed 'The Brick Man' it was another of Gormley's concepts. However, political caution on the part of Labour leaders eventually led to it being denied planning permission in the face of trenchant opposition from the local press and Conservatives. It was also opposed by some who might otherwise have lent support to such a project because they felt it inappropriate for a city still traumatised by the serial murders of the Yorkshire Ripper. As Sandle (2004: 31) recounts, despite the potential for constructing an icon that might stimulate regeneration, political expedience saw council leaders conceding to the 'formidable commonsense of the Leeds public' in the shape of a poll conducted by the Yorkshire Evening Post which had been campaigning against the sculpture.

Left in the Shade: Absences in Leeds' Cultural Policy

There were some cultural dimensions that seemed underplayed by our key decision-makers. The Cultural Strategy emphasises that culture is a broad church. However, for all that there has been large-scale investment in conserving built heritage[4] and that Leeds has more libraries than Birmingham and Manchester put together as well as the biggest free events programme in the country, insofar as a collective mental map exists of cultural policy it is dominated by the arts. Other elements are left in the shade, largely absent from broader discussions about cultural policy in the city. These are: the role of sport and of the creative industries, the diversity of cultural tastes among the Leeds population and the significance of the student market. Here we want to consider each in turn.

Sport   Leeds is home to a formerly highly successful football club, both codes of rugby, a county cricket club that hosts test matches, a (horse) race course, an aquatics centre of international standard and an extensive network of local and regional sport facilities. In our interviews however, apart from respondents with a specific sporting role, the role of sport in the city or its contribution to Leeds' cultural life was rarely considered. Indeed, it required direct questions from us to encourage most of those we interviewed to consider sport's contribution. Those that did talk about sport did so in ways that highlighted its absence from debate: *The thing we haven't mentioned is sport, which is in my opinion an integral part of the culture of the city although it hasn't ever been seen as such.* There is no doubt that sport is featured in the Cultural Strategy, nor that the local authority invests considerable time and financial resources in sport, but it still does not occupy a salient position within most mindsets addressing 'culture'. One respondent, concerned that the council had undervalued the role sport could play in the city, commented: *Sport plays such a prominent part in the lives of so many Leeds citizens, but when I was invited to participate in the Leeds Initiative my initial observation was, "Where does sport feature?"* Others involved in sport have expressed similar concern, but the same might be said by advocates of many other components of culture in Leeds. Passionate advocates are rarely likely to be satisfied with what they have. While other local authorities have sought to attract major sporting events as part of their re-branding and re-imaging, Leeds does not have a record of doing so and recognises it is unlikely to attract the very biggest. However, hosting the 2008 World Corporate Games was designed not only to bring direct economic benefits,

---

4   Not to mention the major heritage attractions of Kirkstall Abbey, Harewood House, Lotherton Hall and Temple Newsam.

but would also represent a showcase with companies like IBM and Anderson's. It therefore represents quite an astute move in addressing the first and third goals of the Vision for Leeds.

Creative industries   Despite Leeds hosting the National Creative Industries Conference in December 2007, and the digital and creative industries identified as one of four clusters to provide the focus for the City Growth initiative in Leeds, the creative industries received few mentions in our interviews. This is in line with findings from research commissioned by the city council:

> The City is widely perceived from within the sector to have not taken the opportunity to mobilise its creative and cultural resources in the most effective way. Indeed, the sector perceives itself to be relatively marginal within the traditional civic and business narratives that have been used to describe and promote the City. (Taylor and Heathcote 2005: 4)

That message came partly as a consequence of comparison with earlier research (Taylor and Heathcote 2004) and observing that despite rapid growth in employment through the 1990s, job numbers had subsequently stalled at around 12,000. Nonetheless, this is a substantial segment of the local economy. As one local policy maker commented: *Employment in the creative industries in the city is at the same level as construction and only marginally below financial services.*
    While many local authorities have striven consciously to develop creative industries, their growth in Leeds has been almost accidental.

> *Leeds is the creative industries capital of Yorkshire, but you wouldn't really know that. A lot of areas like Kirklees and Sheffield make more of a song and dance about it. We dwarf what goes on in Kirklees in terms of cultural and creative industries. In Leeds we've never had a great deal of public sector investment in nurturing and developing creative industries – the success of creative industries in Leeds is solely market led and it has done remarkably well without public intervention.* [local policymaker]

However, if a council takes as its spending priorities those challenges that address the worst problems, demonstrating that the sector is vibrant is something of a double-edged sword.

Cultural Diversity   More generally, the responses of our interviewees seemed to be underpinned by a presumption of a greater homogeneity of 'taste' (Bourdieu 1984) than is evident among Leeds residents. Some respondents could see that the reflection in the policy-making mirror of the need to recognise different tastes is the concern not to offer the poor second best. If questioned directly, no doubt most would insist on the importance of responding to the diversity of the Leeds population, but it is all too easy to overlook this when putting forward arguments around personal interests and expertise. Clearly there are exceptions. One councillor representing an area of cultural diversity observed:

> *Coming from where I do I see the diversity of different groups of people coming into the city and bringing to the city their own culture ... All those organisations that I've mentioned have broadened out and they practised community cohesion before it was called that. We need to be aware that as a city we mustn't marginalise those groups any further but be ready to take them on and to develop closer links with them and share in the cultural life that they have to offer.*

The presumed homogeneity though was more general than a lack of appreciation of cultural diversity. This was made obvious by individual references to different generations or class:

> *Leeds cultural policy is very much about middle class arts ... you need to be part of what's going on to get the benefit ... if you're not part of it it's intimidating getting in the door.* [opposition councillor]

Students   The universities were sometimes recognised as partners in shaping the city's provision, but there was no recognition of the significance of more than 50,000 full-time higher education students and almost 16,000 more part-time students as consumers and producers of culture. Their contribution to club culture and the night-time economy has been considerable and many graduates have moved into the creative industries. Leeds Metropolitan University provides a studio theatre, gallery and music venue that may be principally used by students, but certainly not exclusively. Leeds Metropolitan has also established partnerships with many sporting and artistic clubs, companies and organisations in the city and its region in recognition of the part it has to play in the city's culture and the benefits it derives from that.[5]

What the Cultural Partnership presented as its cultural strategy is not a strategy in the sense of a structured procedure for achieving a set of goals, it is more a statement of celebration, belief and aspiration. For all that there is a strategy, implementation of policy has been incrementalist rather than rationalist (Blume 1979). This is what Lindblom (1968) referred to as 'disjointed incrementalism' characterised by a long process of negotiating/wrestling between the agendas of different interest groups.

Any reading of policy literature suggests we should not be surprised that there is a distinction between strategy as written and strategy internalised in the minds, or the 'assumptive worlds', of key policy figures. Individuals operate with a partial interpretation focussed around personal interests. Reassuringly, it was the two

---

5   This list includes: Yorkshire County Cricket Club, Leeds Rugby, Leeds United Ladies Football Club, Northern Ballet Theatre, West Yorkshire Playhouse, Leeds Art Gallery, Leeds Film, Festival Republic (organisers of Leeds Festival) and several others in the 'city region'.

most senior officers who demonstrated the most complete picture of the strategy's compass.

Cultural Decision-making in Leeds

> In local government we don't spend a lot of time agonising over policy per se. [senior officer]

> There's often talk about cultural policies and I always think it's a bit of a joke. My experience is that there's no such thing. (local councillor)

Cultural decision-making on this scale is clearly complex. By its nature it cuts across many fields of activity. In Leeds, it is not something that commands high priority in any of the political groups of the council. At least it is not an overtly partisan issue – the current administration seems happy to assume credit for decisions taken by its predecessors. As one respondent observed, *Leeds is not an easy city to run*. This comes from its diversity and parochialism:

> *There's the city centre and its periphery and then you have the rim with its deprived inner-city areas and then you're out into either prosperous suburbs or village Leeds and all these places have a feeling that they're independent and they want to have money put into their activities.*

After local government reorganisation in 1974 Leeds was one of the first local authorities to adopt a unitary leisure services department, so the possibility of a coherent cultural policy was there. The opening quotes to this section indicate that such coherence has never quite materialised. Rather, according to the senior officer quoted at the beginning of this section, what appears to have emerged is, *a collection of vision statements*, and the councillor quoted there talked of imprinting personal views of culture and cultural development as corporate policy. In similar vein, another respondent observed: *The first difficulty is knowing whether in fact the city actually has a cultural policy at all*. In trying to establish an underpinning rationale for cultural policy another snorted: *Rationale!!!! You must be joking!* It seems that even with a written strategy there is a disjuncture between the appearance of formalism and 'policy' as practised.

A recurrent theme in our interviews was one of visionary leaders set alongside concerns about lack of interest in cultural issues from most council members. This idea of the visionary leader underlies the development of Leeds' cultural policy. The 'leaders' referred to were, at various times, strong political leaders of the council with an intuitive feel for what is 'good for Leeds', individual councillors with a passion for culture or highly influential local authority officers. The importance of such leaders (or cultural intermediaries) has been crucial in a climate in which few councillors have carried a torch for cultural matters and the Labour group (the largest single party since 1980) felt there were more pressing concerns in social care, housing and education. Even powerful political figures found it hard to push through cultural projects. However, in such circumstances doing deals in pursuit of what was perceived to be good for Leeds sometimes overcame a suspicion of culture and a fiscal cautiousness that was wary of the extravagant gesture.

We found general agreement on the longstanding Leeds way of doing cultural policy, summed-up by one respondent as:

> [an] *idiosyncratic kind of approach dependent on people's personal interests and passions to an extent that was just extraordinary in Leeds and far more than any other big city that I've ever come across.* [policy partner]

Over an extended period then, cultural policy was characterised by opportunism – partly because key politicians did not want their hands tied should the chance for a glamorous new initiative arise; partly because there was little perceived need for a strategy for an area that commanded relatively little priority in the council (Strange 1996). However, the frustrations of officers combined with external pressures from the Department for Culture Media and Sport, the Lottery, the Museums and other Councils finally to produce a cultural strategy in 2002 after a city-wide consultation. This was seen to be crucial in releasing money from external funding bodies. Some in Leeds now appreciate the advantages of having the cultural strategy, while others are still resistant. Despite protestations of the usefulness of the cultural strategy, we have already noted a nervousness on the part of respondents concerned that their knowledge of it might be found wanting, and there appear to be no measures taken to formulate a strategy for the period beyond 2007. This is reminiscent of the rather piecemeal approach to cultural policy before 2000 with its hallmark of creeping incrementalism.

Cultural Partnership

'Partnership' is now so commonplace that it is easy to forget that this has only recently become a favoured approach (Bailey 1995). Other delivery agents were viewed with suspicion. As the leader of the Labour group noted:

> It wasn't just municipal socialism that was unwilling to share a table with others outside. There was a feeling that councils could do it better.

The role of the Leeds Initiative in promoting partnerships was evidenced in this area with the establishment of the Leeds Cultural Partnership in 2002 with the initial task of producing the cultural strategy and then overseeing its implementation. One senior officer commented about this development: *We've set our stall out to develop a partnership approach where the strategy is owned by a body independent of Leeds City Council.* The introduction of this local corporatist approach[6] was an attempt to bridge departmental boundaries within the council and to engage other interests in the city (e.g. Leeds Chamber of Commerce, Leeds Property Forum, the universities/colleges, voluntary groups and other arts, heritage and sports bodies). Given what has already been said about the low level of priority afforded cultural policy it is not surprising that those responsible embraced the opportunity to recruit support from elsewhere. One might reasonably ask what can be achieved by a group of disparate voices that meets once a quarter, but the establishment of the cultural partnership was more about developing and sharing an agenda so that cultural policy was not ghettoised, than about direct implementation and delivery. It also raises questions, though, about the locus of decision-making, giving the appearance that crucial political decisions are being removed from the democratic arena. Even for those directly involved it can be confusing about where decisions are made:

> The external reference group didn't take a decision to say "we're going to settle for the existing building". The council in effect said "Well it's unaffordable to build a brand new art gallery so let's make the best of what we've got" and one way or another that has become seemingly the decision although it was never formally taken by any external consultative body. You have a sort of consultative process and then maybe the group doesn't meet for six months and one day you go along to another meeting and "This is what's happening". So it's a slightly mystifying process but basically in terms of what's happened at the art gallery nobody's going to complain about what they've done because it is very good. But the big issue of should we build a brand new art gallery has been ducked.
> [policy partner]

A similar issue about decisions being removed from the democratic arena arises with the establishment of trusts. One of these was established some time ago to run the Grand Theatre and a sports trust (Leeds Active) is currently being set-up to run sports facilities. It is argued that this confers the ability to *operate in a more business-like manner*, but also *carries the danger of the trusts becoming self-perpetuating elites.*

To most of us the site of decision-making and the source of power seem to lie elsewhere; our respondents were no different. As one councillor suggested: *The average councillor like myself has virtually no power or authority over sport and culture.* That councillor was aggrieved that the Cabinet-style government introduced around the millennium in the interests of more efficient decision-making had served to centralise power. Although it was generally accepted that the

---

6   The reference here is not to Marxist accounts of collective consumption and corporatism in the urban planning of the 1970s but to the introduction of partnership as part of the neo-liberal project in the 1990s intent on introducing more of the market into the welfare state.

Cabinet system had been accompanied by a shift of influence to officers, officers still recognised they had to win support from politicians. When asked how this happened, one senior officer commented:

> Members are canny individuals. They know the things that matter to their electorate. They also care about the image and profile of their city and they know that good quality cultural product promotes the feel good factor for Leeds residents and selling ourselves externally.

Systems and networks providing regular access to the decision-making table were valued, but so too were external allies recognised as 'movers and shakers'.

The Leeds approach to cultural policy might easily be presented as the lack of cultural imagination of both residents and decision-makers of a northern city (and it is not difficult to find elements of that in the records of the time), but clearly things were not quite so simple. Although respondents recognised a lack of enthusiasm among Leeds councillors for cultural projects (not a partisan issue), a Conservative-led council established English National Opera North and took a struggling Grand Theatre into municipal ownership, while a Labour-dominated council oversaw the funding of major cultural projects and attracting Northern Ballet Theatre and the Royal Armouries to Leeds. Similarly, the council has been criticised for not doing more to support its professional sports clubs. However it was (albeit belatedly) instrumental in Yorkshire County Cricket Club securing ownership of its ground and the 20 year agreement for staging test matches. It had also come to the rescue of Leeds United during an earlier financial crisis by buying the ground and renting it back to the club.

Nonetheless, the inability to make a decision or get a decision was a recurring theme. As Bachrach and Baratz (1970) explained a long time ago, non-decision making can be a powerful tool in ensuring some initiatives are not realised, but it can also reflect simple confusion or indecisiveness. The distinction here is between a group able to exercise power by setting the agenda for decision-making and a pluralist system of shifting alliances and indecision. Initially the frustration was not being able to get a (favourable) decision from 'the committee' or for the committee to get a decision from the elite of the ruling group. For all the supposed streamlining and the concern of some that decisions have been removed from the democratic process, similar frustrations remain. From individuals to partnerships, forums to council or board decisions, there are plenty of virtual corridors in which presumed decisions may get lost or delayed, yet non-decisions constitute decisions.

Inward Looking and Outward Facing

We have already made reference to processes internal to Leeds, 'the Leeds way', and to the needs of Leeds, and also to the external pressures of government requirements. In addition there are responses to wider social forces that might arise from economic fluctuations, demographic change, evolving tastes, current preoccupations (such as that with obesity) or major national cultural projects like the Olympics. No city is an island. So, to what extent is a city's cultural policy the product of internal decisions or wider external forces? The answer is of course that it is the product of a complex interplay of both local and extra-local influences and Straw (2004) suggests that local 'cultural scenes' help retain a city's distinctiveness in the face of global cultural forces. Most of our respondents recognised the complexity of this interplay and the often unusual and/or unintended consequences of this interaction of the local and extra-local. When asked if Leeds' cultural policy reflected national moves one regional officer suggested: *It might be nice if it didn't in a way, but I suspect that it does; I think it fits with a contemporary mix.* Asked if they were responsive to initiatives from DCMS, the Director of Learning and Leisure that although DCMS covers culture in all its guises, none of the department's funding comes from there directly:

> *All our money comes through the Department for Communities and Local Government. So we aren't ever in a situation where we're driven by DCMS policy. On a day-to-day basis they don't have an opportunity to influence how we provide our library service or our sport service or anything other than the rules and regulations around CPA [Comprehensive Performance Assessment] that they and the Audit Commission use to measure our success. We happen to have good relationships with DCMS, but I don't hang on every word that comes from DCMS.*

DCMS though is influential in shaping Lottery and PFI applications. The desire to attract external money for major capital projects had been a major factor in producing a formal cultural strategy after 2000, but the Director insisted:

> *I couldn't give you an example of anything in Leeds that we've created with the assistance of Lottery funds that hasn't grown quite naturally from a desire in Leeds to add something or improve something that we've already got.*

On the other hand he was also pleased that Lottery funding had levered more money out of the council for leisure.

Around the UK cultural policy in different places has been directed to (some mix of) city marketing, economic regeneration and job creation, improving quality of life, and raising civic pride (Bassett 1993, Bianchini and Parkinson 1993). Cultural policy is rarely seen in terms of promoting the intrinsic benefits of cultural forms. The development of cultural policy in Leeds is no exception to this pattern. For example, Leeds' bid for one of the large casino licences was represented largely in terms of economic regeneration and employment creation. Beyond that, though, Leeds City Council has an arts and regeneration unit whose remit encompasses more than economic regeneration, moving towards a broader development of the arts. The current city slogan, *Leeds Live It Love It*, represents Leeds' attempt to tackle quality of life issues and raising the profile of the city (the third and fourth of the functions of cultural policy identified above). In one sense this reflects the eschewing of grand gesture policies designed to impress outsiders. However it is undoubtedly common for the cultural offer to be seen in terms of city positioning.[7] One of the propositions is that the cultural offer plays a part in attracting investors to locate in the city, key workers to move here and local talent (including graduates of the universities and colleges) to stay after they have completed their courses. Yet typically our respondents observed that Leeds fails to make the most of what it has to offer in this respect. The unarticulated conclusion (of our respondents) to be drawn from this is that Leeds' cultural policy is failing in this role if the 'cultural product' goes largely unrecognised.

Of particular interest to us is the way in which our respondents made constant comparison between Leeds and other cities. Our informants from the policy-making circles represented four positions when assessing Leeds' cultural offer in relation to other places, though it was not unusual for them to slip between these during the course of the interview. These can be characterised as:

- Leeds compares unfavourably with comparator cities
- Leeds is among the best culturally
- The cultural offer of Leeds is excellent, but has not been promoted effectively
- There may be some aspects of cultural provision/activity in other cities that are ahead of Leeds, but across the full range there are few places that can compare.

The first of these was the most common initial response, though sometimes qualified by an observation that other cities were recognised for their use of culture-led regeneration because their economic and social deterioration had been so much greater than was the case in Leeds. With a more diverse economy Leeds had less of '*a hole to be filled*', and hence neither had to put as much effort into cultural regeneration, nor had such a stark backdrop against which to profile it. Because Leeds faced less dire economic circumstances it had less access than other places to major public funds from national and European government to fund showpiece provision.[8] Others suggested that this success narrative was sometimes used as an excuse for an essentially defeatist, conservative approach with policy-makers content to manage what already existed.

Equally important is how many of our respondents articulated the way in which Leeds is positioned culturally in relation to its close neighbours. This was particularly the case when thinking about how Leeds lacked the cache or benefit of being associated with a recognised cultural success close by. Clearly it is hard to compare like with like, but, for example, many people associate the Sage and Baltic

---

7   The number of times this occurred unbidden persuades us it is not just attributable to our line of questioning.

8   Some see this as too ready an excuse for what they see as lack of response to need.

Exchange with Newcastle even though both are in Gateshead and the Imperial War Museum North, the Lowry and Old Trafford (home of Manchester United and Lancashire County Cricket Club) with Manchester even though they are (variously) in Trafford and Salford. And the BBC's move that is typically referred to as being to Manchester is to Salford. However, few would see Bradford's National Media Museum or Harrogate's International Centre as part of Leeds.

According to the council's web site the Leeds city region has a population of 2.8 million and encompasses the surrounding districts of Barnsley, Bradford, Calderdale, Craven, Harrogate, Kirklees, Selby, Wakefield and York. Whether they all recognise that is another matter. The surrounding local authorities have been reluctant to cede to Leeds the kind of pre-eminence assumed by Manchester among its neighbours. But it is not a one-way relationship. As the leader of the Labour Group observed: *There's a long way to go to get Leeds to think outside Leeds and to the city region.* Perhaps at the instigation of Yorkshire Forward (the Regional Development Agency), there appears to be greater regional co-ordination. Leeds was prevailed upon not to contest Bradford's bid to be European City of Culture and there is recognition of the need for Leeds not to compete with neighbouring cities to be the host for the same national teams pre-Olympics. It may be that the programme for the cultural Olympics proves to be more significant for cities like Leeds than the sporting event. In line with the market forces operating in many other sectors of the economy, Leeds has (along with Sheffield) become a regional hub for the growing cultural industries (Bell *et al.* 2007: 5). The status of regional capital and expectations of what goes with that meant some of the policymakers, despite personal reservations, had felt obliged to bid for one of the large casino licences offered by the government, believing it was what people would expect of a major city.

Some of our interviewees criticised 'Leeds' for wanting cultural events or infrastructure simply because other cities had them or were developing them. For example, Leeds had no national museum, so acquired the Royal Armouries; had no ballet company so acquired Northern Ballet Theatre. Leeds has been without a major indoor venue capable of hosting a range of different events for a number of years, something that is often seen as a weakness. However, the debate about the need for a new venue only acquired real urgency when comparisons started to be made with other regional capitals with large multi-purpose venues.[9] We must be wary of resorting to trite explanations based on Yorkshire competitive spirit (after all you don't need to compete when you 'know' you are as good as anyone else anyway); the explanation lies instead in a feeling of doing right by Leeds. If other people have something desirable why should the people of Leeds not have something as good? Good cultural provision (typically referred to in current terminology as the cultural offer) is required not just in terms of local consumption, but also in supporting the city's economic function.

---

9   Plans for a large-scale arena have now reached the stage where SMG has been named as the preferred operator.

Change and Continuity

There have been key figureheads in terms of cultural policy who have come and gone; sometimes councillors, sometimes officers, sometimes those working in cultural organisations. Some have moved on to greater things, but a pragmatic organisational culture and more recently having to articulate a cultural strategy, have ensured some measure of continuity. However, the majority of our respondents did feel that the orientation to cultural policy in Leeds had changed over the past five to ten years. This was variously attributed to organisational change, a different climate of ideas and the influence of key individuals. Nationwide there are greater expectations of what can be achieved through cultural activity and these are reflected in the *Cultural Strategy for Leeds*.

Pressure from the New Labour government not only required cultural strategies, but also a reorganisation of local government so that around the millennium a cabinet style administration was introduced in Leeds. Still viewed with suspicion by some for removing decisions from council debate, others saw it as a positive move towards *a more strategic view of how it was going to develop its capital projects, how they would relate to each other, what order they would apply for the money and so on* [policy partner].

At a time when culture is increasingly delivered electronically, (public) cultural policy is still preoccupied with museums, galleries and theatres (and sometimes stadiums and parks) in the longstanding tradition of civic provision. The written document that is the *Cultural Strategy* also reflects the wealth of sports development and arts activity that has built up over the years and draws attention to 'the mixed cultural economy of public, private, voluntary and community interests' (Leeds Cultural partnership 2002: 14). Undoubtedly shifts have occurred, but this more expansive conception of cultural policy has a precarious foothold.

We can also track the consequences of organisational change. The change to cabinet governance in Leeds in 2000, with its more corporatist approach to policy-making, shifted the balance of power from councillors to officials. We have also identified another element of a more locally corporatist approach in policy shifts that have leant greater significance to partnerships between the different sectors with a view to reaching beyond the council and engaging other key players in cultural provision. Designed to involve more and encourage them to recognise their potential contribution, this shift to partnership has also left some feeling they have even less power to influence the city's culture. Organisational change administratively has also affected the direction of policy as delivered. For example, when responsibility for schooling was assigned to Education Leeds, matters of lifelong learning were aligned with leisure such that the discourse of officers incorporated ideas of leisure as skills development.

Seeking to promote change the *Cultural Strategy* has also conferred stability during other changes. To date change of political control of the local authority in 2004 appears to have had little impact on cultural policy: '*the change of administration has not rocked the boat... There's a surprising unanimity across political groups about the value and role of culture and leisure'*. This is certainly true in terms of major financial projects, as the decisions regarding investments coming to fruition were taken under the previous administration. Successor projects, however, have been slow to appear. Implementing plans for the arena and for the large casino licence awarded by the government's Casino Advisory Panel is proving a drawn-out process and the establishment of the sports trust has encountered difficulties.

De la Durantayne and Duxbury (2004) maintain that what was previously 'under the radar' is now part of the civic mainstream, offering a sense of place and increasing community attractiveness and quality of life that increase 'place competitiveness'. Our respondents typically observed a wider appreciation of the value of culture now than in even the recent past. We have already pointed to a suspicion of matters cultural in Leeds' political circles. This undoubtedly persists, but nevertheless culture has been moving up the political agenda, perhaps partly because of a loss of powers in other areas. This has combined with a shifting of horizons for the city, which dates from when a previous leader of the council set his stall out to make Leeds a regional capital in European terms. This 'Europeanisation' of Leeds was a vision that the business community could subscribe to and one in which the culture of the city was central. It was also noted that the Regional Development Agency is belatedly giving more credence to the potential of cultural investment.

So, recent years have brought greater awareness of the potential contribution of culture, a more coherent strategy, more effective bidding for external funds, greater collaboration between departments and other partners. However, set against this, while cultural policy may be largely non-partisan, it is not immune from inter-party political point scoring. Moreover, with Lottery funding declining, having passed the end date of the Cultural Strategy, with a decision not to replace the Director of Learning and Leisure as another administrative restructuring was implemented, and without the obvious cultural figureheads of the past, the Cultural Partnership has a further challenge to champion culture across the city.

Conclusion

We look back to give a sense of perspective, an appreciation of how the current position has been reached. It would be unfair to suggest that the policymakers are trapped looking back, even though they are sometimes criticised for not looking forward. It is certainly the case though, that past investments and decisions have a powerful influence on current provision and activity. The oil tanker that is 'the Leeds way' presses on. The story of cultural policy in Leeds is one of gradual accretion with occasional surges. However, with the exception of a possible arena, there seem to be no new facilities on the horizon that were not planned several years ago. Leeds stands accused of being complacent in cultural terms, happy to roll with whatever is generated or attracted by the city's economic success and not being prepared to back its aspirations with financial investment. While proponents of this argument point to all sorts of instances to support their concerns, there has been considerable investment in arts, sport and heritage assets in recent years. In no small part this has been supported by Lottery funding, but the squeeze on those funds effected by the Olympics over coming years will make it harder to raise finances for cultural facilities in Leeds.

One of the intriguing aspects of doing this kind of research is how people are able to interpret the same information differently depending upon their particular interests. For example the consultant's report on venues in Leeds was used to justify building an arena, upgrading the town hall and building a new concert hall. Some consultants' reports may be as vague as the Oracle at Delphi, but some of the conclusions of our respondents required a very particular interpretation of the recommendations. What most did agree on was that Leeds has not capitalised on its cultural offer in marketing terms in the same way as other cities have managed to do; surely a cardinal sin in postmodern times. Nonetheless, Leeds has enjoyed spells as the club capital of the UK and been hyped so much that it is difficult not to imagine it as some type of retail 'centre of excellence'.

We have emphasised the difference between the stress of the Cultural Strategy on the people of Leeds and the magnetic pull for 'policy influentials' of large investment projects. Respondents though still indicated a strong belief that cultural policies can deliver improvements to people's lives, but: '*there has been a bit of over-claiming in cultural policy. I think cultural policy has to see itself in quite a humble way – we are part of the mix*'. Investment in the 'cultural offer' in Leeds has been viewed quizzically or more actively resisted by some on both the political right and left: the former because of a belief in limiting public expenditure; the latter through a desire to prioritise social services (in the broad sense). Without necessarily subscribing to Florida's (2005) notion of the creative class, more and more decision makers in Leeds seem to accept that culture is central to the success of the city. After all, in the UK, cities are unable to offer tax incentives as they do in the United States to attract employers. Whatever the successes of the *Cultural Strategy* may have been, its lifespan ended in 2007 and there are no plans to produce a revised version. Hopes now seem to be pinned on the Cultural Partnership to take forward the cultural elements of the *Vision for Leeds* through building alliances and, in true postmodern fashion, shaping the discourse of what Leeds is to be.

References

Bachrach, P, and Baratz, M. (1970). *Power and Poverty: Theory and Practice*. New York: Oxford University Press.

Bailey, N. (1995). *Partnership Agencies in British Urban Policy*. London: UCL Press.

Bassett K, (1993). Urban Cultural Strategies and Urban Regeneration: a Case Study and Critique. *Environment and Planning A*, 25(12), 1773-88.

Bell, D., O'Connor, J., Taylor, C. and Gonzalez, S. (2007). *Yorkshire Cities and Culture. A Review of Current Thinking*. Leeds: University of Leeds.

Bianchini, F. and Parkinson, M. (eds.) (1993). *Cultural Policy and Urban Regeneration: The West European Experience*. Manchester: Manchester University Press.

Blume, S. (1979). Policy Studies and Social Policy in Britain. *Journal of Social Policy*, 8 (3), 311-34.

Bourdieu, P. (1984). *Distinction: A Social Critique of the Judgement of Taste*. London: Routledge.

Bramham, P., Butterfield, J., Long, J. and Wolsey, C. (1994). Changing Times, Changing Policies, in *Leisure: Modernity, Post-modernity and Lifestyles*, edited by I. Henry. Eastbourne: Leisure Studies Association, 125-34.

Department for Culture, Media and Sport 2000 *Creating Opportunities Guidance for Local Authorities in England on Local Cultural Strategies*. London: DCMS.

de la Durantayne, M. and Duxbury, N. (2004). Cultural Development in Cities: Linking Research, Planning and Practice. *Loisir et Societé*, 27 (2), 320-22.

Florida, R. (2005). *Cities and the Creative Class*. New York: Routledge.

Leeds Initiative (2004). *Vision for Leeds, 2004-2020*. Leeds: Leeds City Council.

Leeds Cultural Partnership (2002). *The Cultural Strategy for Leeds, 2002-2007*. Leeds: Leeds City Council.

Lindblom, C. (1968). *The Policy Making Process*. Englewood Cliffs, NJ: Prentice-Hall.

Maconie, S. (2007). *Pies and Prejudice: in Search of the North*. Ebury: London.

Sandle, D. (2004). The Brick Man Versus The Angel of the North: Public Art as Contested Space, in *Leisure, Media and Visual Culture: Representations and Contestations*, edited by E. Kennedy and A. Thornton. Eastbourne: Leisure Studies Association, 187-202.

Strange, I. (1996). Pragmatism, Opportunity and Entertainment: the Arts, Culture and Urban Regeneration in Leeds, in *Corporate City? Partnership, Participation and Partition in Urban Development in Leeds*, edited by G. Haughton. and C. Williams. Aldershot, Avebury.

Straw, W. (2004). Cultural Scenes. *Loisir et Societé*, 27 (2), 411-22.

Taylor, C. and Heathcote, C. (2004). *The Creative Industries in Leeds: An Initial Economic Impact Assessment*. Leeds: Leeds City Council.

Chapter 5

The History Boy: Made in Leeds

Peter Bramham

AB: Finally a word about what I always consider rather a dubious role, namely 'a Northern writer'. Northern writers like to have it both ways; they set their achievements against the squalor or the imagined squalor of their origins, and gain points for transcendence, while at the same time asserting that somehow Northern life is truer, and in some undefined way, more honest than a life of Southern comfort. 'Look, we have come through,' is the message, but I can't quite see why a childhood in the industrial north is less conducive to writing or whatever, than a childhood in Petersfield or Wimbledon or wherever. I mean it's quite true if you're born in Barnsley and you set your sights on becoming Virginia Woolf, it's not going to be roses all the way. [Laughter]

Alan Bennett was interviewed by Martyn Auty at the National Film Theatre on 19th September 1984.[1]

This chapter explores the tension between local roots and wider metropolitan culture; between an industrial past and present and then sudden transition to a postmodern future. This journey may be mapped against Alan Bennett's own life, his deep childhood roots in Leeds, the Oxbridge scholarship, acting, and on to national and international fame as a playwright and author. Alan Bennett's generation witnessed the dissolution of modern industrial life and has experienced diversity and difference, a distinguishing characteristic of the postmodern condition. But whatever cosmopolitan cultural capital Bennett has acquired, his distinctive national appeal lies in his parochialism, his provincialism and his equivocal relationship to change. It is hard to see Alan Bennett as in any way representative of his generation growing up in Leeds or why his distinctive localism should be translated into the status of 'a national treasure', but whether managed or not, there is little doubt he enjoys iconic media status in Leeds, in Yorkshire and, in what the metropolitan elite would dismiss as, 'the rest of the country'.

How he has managed this life of celebrity is an interesting story in itself. But it is his story and the way it has been revealed, somehow reluctantly and intimately, backing blinkingly into the limelight has attracted intense media interest and approbation. Indeed, Bennett describes himself as a fiercely private person, "one who can scarcely remove his tie without having a police cordon thrown around the building" but, such is the media interest, that though he rarely does interviews he

---

1 See the Guardian Interview 1984.

does publish extensively. Having been described by the Independent on Sunday as 'winsome', he refused his pre-arranged interview by sending a postcard to the paper which simply stated 'Win some, lose some'.[2] However, one needs only go to the web[3] to be overwhelmed by the output of his life's work in drama, fiction, radio drama, screenplay, short stories and, not least, his diaries. One of his latest books includes his diaries from 1995-2004, Untold Stories (2005) and was written on the assumption that it would be published posthumously because his odds of surviving cancer at the time were deemed to be slim. Bennett himself acknowledges that the bulk of his diaries require strict editing as they focus on the mundane, with two thirds of entries dismissed as 'simply boring', a designation assigned by most reviewers to Michael Palin's recently published diaries. But such is the synergy of Alan Bennett's own biography with his published work that any diary entry from this northern celebrity is eagerly awaited after the success of Writing Home (1994).

This public and private tension is first articulated by Bennett (1994) in his book Writing Home (1994) as he writes about his personal struggle to find a literary voice in his life. Indeed, for someone so productive and prolific, he frequently agonises in Untold Stories (2005) over the process of writing, ambivalent about accepting his own identity as a writer, but nevertheless peeved and chastened when one interviewer suggested that he was presently experiencing 'writer's block'. His anger related to journalists naming and labelling complex things but it was a sensitive issue and a time of illness. Bennett admits that writing plays 'ties you up in knots' and 'ruins you for other things in life'. When he first started writing in Oxford his routine was to start with one or two drinks (a self-deprecated limited capacity for manly drinking because of a duodenal ulcer) to loosen the repressions of his local past. Should he write in a cosmopolitan Oxford-educated voice about 'art' or explore his local, respectable working-class background in Northern Leeds? As his Mam and Dad would say to him as a child, it is all a question of 'not putting it on' or 'showing off'. The tension between northern culture and metropolitan culture, between 'relevant' and 'useless' knowledge is the key theme in Bennett's most recent award-winning play, The History Boys. Interestingly, this is a northern play about teaching, or rather coaching, pupils in history for Oxbridge, set in Sheffield in the 1980s, and yet the production still enjoyed massive success with audiences in both London and New York. It has won three Olivier Awards in the UK and six Tonys in USA. But, as so often with Bennett's work, the play operates on many levels. It is a harbinger of debates in education and teaching styles but subversively explores eroticism, homosexuality, adolescence, schooling and the media.

This chapter on Bennett and his city is informed by historical sociology in that it focuses on a particular age cohort of Leeds-based northern intellectuals.

---

2   BBC Radio 4 Archive Front Row with Mark Lawson, January 2008 to November 2007, accessed 15.07.2008.

3   See for instance Author Profile www.contemporarywriters.com.

Postmodern theory would now cast them as cultural intermediaries who articulated and made sense of emerging lifestyles, a unique and differentiating feature of the postmodern condition. Each generation grows up in, experiences and makes sense of, its own historical setting or context, shaped by distinctive economic, political, social and cultural formations. It is useful to our understanding to adopt some kind of historical perspective as outlined by the late Philip Abrams (1981). To understand present urban life one needs to explore the changing collective experiences of generational cohorts and how these are expressed over a lifetime. This requires some Durkheimian analysis which focuses on the '*conscience collective*', that social solidarity which crystallizes at distinctive historical episodes. Durkheim's work on deviance stresses the functions of social rules and rituals, the effervescence and celebration of shared experience and memories. Peer pressure and shared historical experience are crucial contexts and resources for individuals growing up and growing old together, producing a distinctive 'spirit' for every age. Each generation inherits both material and cultural resources from the previous generation and in the lengthening longevity of modernity each generation must endeavour to win some cultural space from its parents and from the past. The challenge for historical sociology is understanding how individual biographies map into generational and long-term structural changes, how universals of class and generational experience are realised in particular local representations and contexts. One distraction of postmodern writing is that it often privileges nuanced accounts of individuals, difference and change whilst ignoring or glossing over collective processes of similarity and continuity.

The cohort that interests us here was born in the 1930s, shaped by pre-war austerity, grew up as teenagers in the post-war austerity of the 1940s and 1950s, but also drew on the 1960s 'expressive revolution' of drugs, sex and rock 'n roll. This transition helps provide them with a distinctive vantage point from which to view historical change. In one sense this cohort is the last of the modern, standing on the threshold of all that is postmodern. Indeed they ushered in the postmodern with their nostalgia for modernity, their celebrity and their prurience. They were subversive through humour, their own sexuality and became northern lights or a northern constellation in the metropolitan arts cultural firmament. Now in their seventies, they were located on the cusp of what has previously been described as the 'ants' and the 'grasshoppers' generation (see Bramham 2005, 2007). The 'ants' parent generation, struggling through years of adversity in the forms of war, poverty and unemployment, finally established the welfare state in the UK. By way of contrast, the next 'grasshopper' post-war youth generation of 'baby boomers' enjoyed the fruits of this public collective investment, only in their turn, as the parent generation, to opt for individual choice, market forces and widening inequalities. To use Bauman's (1992) terminology, two thirds of the population were seduced into consumer culture and postmodern lifestyles whereas a residual bottom third of the population were excluded, restricted to meagre state benefits, constrained by state workfare surveillance. The 'grasshopper' generation, the so-called 'baby boomers' or 1960s generation, resisted the burden of funding collective provision in health care, education and housing and greedily profited from various share issues from the pragmatic privatisation policies of a Thatcherite neo-liberal state (Hutton 1995). Bennett himself agonised about going private to treat his colon cancer in 1997, comparing NHS hospitals favourably with privatised commodified health care. But, as he reluctantly and unsurprisingly acknowledged, the private sector provided immediate surgery and so his money, rather than BUPA insurance, bypassed lengthy public-sector waiting lists. He acknowledged in a BBC interview that it probably cost someone else's life as they quickly found a space on the waiting list.[4] Nevertheless, he has remained one voice on the political left, against the Iraq war, with a life-long commitment to some form of welfarism. In 2008, he suggested that private education in the UK should be banned as it fed class divisions.[5] When fêted by Leeds he affirmed in 2006, "To be a freeman is a great honour and I would like to take it as a testimony to that education the city gave me free of all charge so many years ago." Leeds City Council leader Andrew Carter suggested that the honour was well-deserved and overdue, "While his glittering career has taken him to all corners of the world, he is a west Leeds lad who has never lost touch with his Leeds roots".[6]

This chapter maps out his distinctive Leeds-based voice. It is beyond its scope to explore in detail the provocative, influential and distinctive contributions of other northern intellectuals and literary aesthetes such as Richard Hoggart, or poet Tony Harrison, as well as Bradford-raised, David Hockney. They also constitute that generation's pathfinders from ordinary elementary school, to grammar school, up to Oxbridge and onwards into the metropolitan world of university life, the BBC, art galleries and cultural television. This is no easy journey as real tensions emerge between parental working-class culture and the educated scholarship boy, most clearly in the career of Tony Harrison. His parents were intentionally separated from the publication of 'public poetry' which respectable working-class culture dismissed as 'mucky' or 'dirty' work by addressing taboo issues such as sex in 'foul' everyday language, palpably highlighted when his controversial 'film/poem' V (Harrison 1985), was broadcast by Channel Four on the 4th November 1987. The poem, stimulated by graffiti, daubed by Leeds United fans in his parents' graveyard, reflected on language and class tensions exacerbated by unemployment and the UK miners' strike in 1984. David Hockney also made the journey from Bradford Grammar School to Bradford School of Art and onto the Royal College of Art, winning 'glittering prizes' all the way, before escaping to the sunshine of California, with its long shadows, swimming pools and more relaxed gay scene. Like Alan Bennett he refused the honour a knighthood in 1990 and, when aged seventy plus, he complained bitterly about New Labour's nanny state with its health agenda which has banned smoking in public places, criminalised

---

4　BBC Radio 4 Archive Front Row with Mark Lawson, January 2008 to November 2007 accessed 15.07.2008.

5　See '*Bennett backs private school ban*' BBC NEWS 24.01.2008.

6　See '*City freeman honour for Bennett*' BBC NEWS 21.03.2006.

marijuana use and more recently attempted to regulate computer-generated images of children.

The privileged iconic status of this age cohort offers one prism through which to view significant changes in both Leeds and wider postmodern culture. The chapter explores how this particular cohort articulates its distinctive Northern roots – the unique and often idiographic reminiscences of Richard Hoggart's Hunslet and Alan Bennett's Headingley. These times and places are embedded in urban industrial modernity. They have been mapped out sociologically in classic UK community studies of the time such as Brian Jackson and Dennis Marsden in Huddersfield (Jackson and Marsden 1962), Dennis, Henriques and Slaughter on Featherstone (Dennis, H. et al. 1969), Jeremy Tunstall on Hull (Tunstall 1963) and so on – that nostalgia for the urban white working-class community, its gendered networks, the celebration of its manual work, leisure and culture. These studies sit some distance from middle-class culture and authority, as they document cultural space negotiated by ordinary people in everyday life, clearly articulated and agonised over in later work of the Centre for Contemporary Cultural Studies.

Class and Cultural Change

In fact, Richard Hoggart belongs to an earlier generation, as he grew up in the 1920s, went to Cockburn High School in South Leeds, and graduated before World War II from Leeds University. He was Staff Tutor at Hull University when he published his seminal and iconic work, The Uses of Literacy (1957). This book, originally titled The Uses and Abuses of Literacy (Owen 2005) was primarily autobiographical in nature and documents growing up within a vibrant close-knit working-class culture in Hunslet, South Leeds. He bemoaned the corrosive massification and Americanisation of local UK cultures in the post-war period. His main emphasis in the first half of the book, An 'Older' Order, was on the real world of working-class people and on the entrenched divisions between 'Them, and 'Us', between the working classes and the myriad of people with power – bosses, doctors, public officials and so on. Part Two of the book, Yielding Place to New, traces forces of cultural change leading towards mediated mass hedonism. This loss of 'authenticity' would prove a central research topic for postmodern writers interested in theorising about space and place, such as the work of John Urry (1990, 1995).

At the beginning of the sixties, Richard Hoggart, alongside Raymond Williams (1963), invoked cultural studies to research class and popular culture. His reading of working-class culture generated a rich legacy. By 1964 he had founded the Centre for Contemporary Cultural Studies at Birmingham University (CCCS) and under his tutelage and that of Stuart Hall, the next generation of 1960s 'baby boomer' researchers such as Phil Cohen, Paul Willis, Paul Corrigan, Angela McRobbie, Dick Hebdige and Paul Gilroy explored relationships between class, youth and the mass media. Their common project was to fuse Marxist literary studies, history, sociology and political theory to make sense of changing relationships of class, media and communities. Their primary focus centred on change as their ethnography explored how class fractions of youth developed subcultural styles to create different identities in response to their parental generation, class position and media relations. Rather than 'cultural dopes or dupes' controlled by capitalist structures, cultural studies research categorised youth as 'agency' striking its own local path through changing historical circumstances in the 1960s and 1970s. Class relationships, education, community and mass media became viable topics for academic research and commentary.

People like Us? We are Leeds

Bennett's writing evokes a distinctive sense of both time and place. He was born in 1934, educated at the same Upper Armley Christ Church Elementary School as Barbara Taylor Bradford, the Yorkshire romantic novelist, between 1939 to 45, though briefly evacuated during the Second World War, and from 1945-1952 he was at the local state grammar school, Leeds Modern School. After two years National Service in Cambridge he eventually won a scholarship to Exeter College, Oxford, gained a first-class degree (just, by his own account) and became a researcher and lecturer at Oxford University. He is at his most vivid when describing industrial Leeds where he grew up...places such as Gilpin Place, Tong Road, Stanningley cinema, Woodhouse Ridge and Leeds Town Hall and City Library. The picture painted mirrors Lowry's Salford – it is full of terraced housing, canals, mills, trams, railways, with the River Aire its centrepiece – as black, steaming, polluted. It leaves the Yorkshire Dales in Malham Cove crystal clear but after travelling through Keighley, Bingley, Saltaire, and Shipley carries all the marks and smells of factories and manufacturing of the industrial north. The language and phrases too are all Leeds Loiners'[7] talk: there are 'ginnels' (for alleyways), 'roaring' (for crying), 'spother' (for fuss), 'dolled up' (for dressed up to go out), 'traipsing' (for travelling) and distinctive local phrases: 'that's put the tin hat on it', 'you've got this place like a palace' and in recorded conversation, such as between a policeman and prying spinster, Miss Ruddock: was it me who had been writing these letters? I said, 'What letters? I don't write letters.' He said, 'Letters. I said, 'Everyone writes letters. I bet you write letters. He said, 'Not like you, love" I said. "Don't love me. You'd better give me your name and number. I intend to write to your superintendent' (Bennett 1988).

Bennett comes from a respectable lower middle-class family, although he assigns himself to the working class, alongside Tony Harrison who he describes as attending a 'posher' grammar school and so under even more pressure to lose his working-class accent. Alan Bennett's dad was a reluctantly apprenticed butcher at

---

7   "Loiner" refers to the inhabitants of Leeds but its precise origin is uncertain see http://www.leeds.gov.uk/About_Leeds/History.aspx accessed 27.07.2008.

the age of 14 years due to the tyrannical wishes of a step mother, nicknamed in the Bennett household as an old 'Gimmer' (a sheep with no lambs). He eventually ran his own business next to the tram sheds in Headingley. As Leeds has remained a class-divided city, moving from poor West Leeds to the more affluent North of the river was no mean feat. There is no mention of his mother's occupation but Mam's province was the domestic sphere, with its purity and pollution rituals and routines, with her neat hierarchy of buckets, cloths and mops to deal with the dust and grime of Leeds in those days. Consequently, there was a traditional sexual division of labour in the Bennett household with his father the breadwinner and continuing a wide variety of leisure pursuits and fads such as playing the violin and double bass, fretsaw toys, and riding a motorcycle and sidecar.

Bernice Martin (1981) has argued both respectable working-class and middle-class culture offered a strong set of values, codes, rituals and routines, so as to differentiate themselves from 'rough' working class fractions. Following Mary Douglas' key concepts of 'group/grid' and 'ritual/zerostructure,' also reflected in Basil Bernstein's work on elaborated and restricted linguistic codes, she argued that strong group and grid means strong symbolic boundaries and distinctive rituals. These express social control and internal divisions, whereas weak group and grid provide for the 'conditions for effervescence' for cultural liminality, religious ecstasy, chaos and individualism. Martin writes about working-class industrial Lancashire drawing on the work of Roberts (1971), Seabrook (1982) and Pearson (1976) which stresses the importance of collective solidarity, with clear boundaries drawn between 'us' and 'them'. Traditional working-class culture, with its strong grid, abhors ambiguity and difference so that homosexuality and black ethnicities straddle and subvert traditional categories. Alan Bennett's childhood and family members were ingrained with this rigid working-class culture and his life and writing playfully deconstruct many bogus certainties that shaped modernity.

Alan Bennett's Childhood

> That is by the way, but then so is much of this reminiscence, my childhood itself fairly by the way, or so it seemed at the time. Brought up in the provinces in the forties and fifties one learned early the valuable lesson that life is generally something that happens elsewhere. Bennett (1988: 13)

Although Bennett avoids being stigmatised as 'one of them', castigated by a repressive working-class culture that vilifies homosexuality, 'the love that dare not speak its name', he nevertheless struggles to feel at home inside his family and his grammar school, but during adolescence finds some solace in rigorously immersing himself in the established routines of the Anglican church. He adopts Christian tradition with fervour but even this devotion carries for him hidden risks. At sparsely attended eucharists at St. Michael's Headingley the young Bennett worries that the chalice may carry diseases such as TB and cancer. At busy services such as Easter and Christmas, with a church replete with publicans and sinners he agonises about catching VD from the chalice from the mouths of casual churchgoers.

> That I should catch syphilis from the chalice might be all part of His plan. The other place I was frightened of contracting it was the seat of a public lavatory, and that the rim of the toilet should be thus linked with the rim of the chalice was also part of the wonderful mystery of God. It was on such questions of hygiene rather than any of theology that my faith cuts its teeth. (Bennett 1987: 13)

But Bennett's position is both detached and reflective. Throughout adolescence he feels strange, isolated, different from everyone else, queer and separated from ordinary lives. He feels this acutely in his early childhood with his Mam and Dad, Lillian and Walter, not able to 'mix', feeling ill at ease with the less respectable family relations of Aunties and Uncles who seem to be more relaxed, drink alcohol, enjoy holidays and generally have more fun and sex. The Bennett family do have aspirations to 'mix' and fit in with their neighbourhood but, whether in urban Leeds or the Dales village of Clapham, they do not manage it and often feel under the gaze of critical eyes and possibly shaming judgements. Bennett deals with this with tongue in cheek suggesting that Lillian's purchasing of peanuts and cocktail sticks is perceived as the royal route to successful socialising and partying. But his diaries also provide shocking and graphic details of the mental anguish and depression that tormented his mother, his aunty and clearly became the miserable lot of many of that generation of women who struggled to be housewives and homemakers. In adolescence Bennett feels that his life is on hold, his delayed puberty a painful stigmata written into his body – as, in Goffmanesque fashion, he daily surveys his peer group at school for those tell-tale signs of puberty. He is tormented by his own boyish looks, the unbroken voice, not shaving, the donnish spectacles and so it goes on into his twenties and thirties. Embarrassment and shaming are central processes in working-class culture with its repressive solidarity and Bennett is more than sensitive to its pressure. Normality appeared to be beyond the scope of the Bennett household and family, adulthood often seemed a step too far for the young Alan Bennett. He was 26 years of age when he first appeared in Edinburgh footlights, with Beyond the Fringe, and he always felt that his undergraduate colleague performers i.e. Jonathan Miller, Peter Cook and Dudley Moore were both much younger but also much more worldly and sexually mature than he. On being asked the question by Sir Ian McKellen, in 1997 whether he was gay, he enigmatically replied, "That's a bit like asking a man crawling across the Sahara whether he would prefer Perrier or Malvern water" (2004).

When his sexual life did eventually take off in London, he found himself reluctantly drawn back to visit his mother in the North during her bouts of depression and subsequent institutionalisation in mental hospitals. His train journeys between London and the North became a recurrent feature, both literally and metaphorically in his lifetime. He is acutely aware that he has an unusual constituency, a distinctive public for his work. Sometimes he addresses them as a metropolitan elite as in his prefaces to plays, about art, in literary debates for the cultural giants that stalk the papers of the London Review of Books, amongst notes about theatrical and television productions, whilst simultaneously he is acutely aware of his Northern constituency or public when reading or performing. There is always some pressure out there, as clearly illustrated when reading one of his latest works at the West Yorkshire Playhouse, he 'was accosted on the way in by two sabre-toothed pensioners: 'It had better be good', warned one of them. 'We are big fans of yours' (Bennett, 2005, 386).

The cultural bedrock of ordinary people functions not so much to keep chaos at bay but rather to provide distinctive landmarks by which people orientate themselves to know where they are. This distaste in the respectable working classes expresses itself as loathing of people who were 'common'. Bennett's mother and he himself hold on to strong views about what constitutes 'commonness' in values, opinions and behaviour. However, Bennett ambivalently remembers the vulgarity of his mother's unmarried sisters, Aunties who were career girls, or as Bennett has it, shop assistants. His elder brother, Gordon, joined the RAF as a pilot, married, had children and grandchildren and would feature centrally as the nearest relative (living in Bristol) when Lillian Bennett was in her final residential care home in Western-super-Mare, suffering from dementia, after struggling with decades of depression. Interestingly, it is his mother's side that sets the tone for the family, as Granddad Peel was originally from a wealthy manufacturing middle-class background and, unbeknownst to Alan until 40 years on, committed suicide by drowning in the Leeds-Liverpool canal at Rodley. Families struggle to keep secrets and over the years, Bennett eventually researched the death, reported in an article in the local newspaper. Gaining an old Ordinance Survey map, he discovered the precise spot of his Grandad's suicide. This corner of Leeds becomes a significant place for him, rather than a green space previously ignored on home visits whilst regularly commuting on the Leeds-London train.

Education and Schooling

Alan Bennett is living testimony to a meritocracy that more or less underpins the post-war settlement of welfarism. It was a central role of the local state to provide educational access and opportunities free of cost to the next generation irrespective of background. This was a political and cultural ideal fought for by the 'ants' generation in the 1930s who were themselves denied secondary education, with the exception of a minority elite who were privileged to 'stay on' and then take up very few university places. Alan Bennett's parents were typical of their generation in that they saw education as the means to social mobility and 'bettering oneself'. Indeed, public libraries, mechanics institutes and the Workers' Education Association were organisations expected to lift aspirations and empower ordinary people. Walter and Lillian Bennett felt it was precisely their lack of education that ensured that they did not 'mix' and that their bright younger son would face few obstacles 'in that department' because of his educational experience and expertise. There is some residue in this when Alan Bennett talks over and for his father when dealing with the Settle doctor who was treating his mother's depression in the 1990s. Yet Bennett himself ironically points out that he was unaware of family secrets and particular skeletons in the Peel family's cupboard such as suicide and mental illness.

So a major part of Bennett's identity is of a Leeds state-school scholarship boy. He remembers nostalgically the solid coat of arms, with its owls and the lamb in a sling, inscribed *Pro Rege and Lege*, For King and the law, proudly symbolising a Labour-led City Council on its trams and buses, exercise books and so on (Bennett 2004: 513). When going up to Oxford he felt himself to be representative of or ambassador for Leeds and the grammar school system with its deep-held tradition of free education and public access. Leeds was embedded in his adolescent persona as he spent his evenings friendless, as a flâneur, roaming suburban streets, absorbing sights of different people and places. Whereas postmodern Leeds boasts city-centre shopping and heritage destinations as sites for the tourist gaze, for Alan Bennett it was mostly local churches and parochical streets, although he has much to say on Leeds Art Gallery, about the Reference Library next to the Town Hall and also on County and Thornton Arcades. But he is uncomfortable about changing Leeds, about the wilful destruction of Victorian architecture in the 1960s and 1970s and even more so, when in 2003, his Leeds Modern School is demolished, replaced by a postmodern building which could to his mind be a factory or shopping outlet. It is interesting that he knew little of the redevelopment of his old school but took the opportunity as a celebrity, when invited to appear on Yorkshire TV's 'Calendar News', to comment on his schooldays memories. No doubt he would welcome the recent refurbishment and opening up of the tiled ceiling in the Arts Library but possibly bemoan its new use as a postmodern café. That said, he readily acknowledges that people go into libraries and art galleries for a wide variety of reasons, not necessarily to read books or see works of art as a primary function. Even his valued Leeds City Education Committee, a beacon of modernity and welfarism has recently been taken out of local authority control and privatised, run by an agency, branded Education Leeds.

Much of *Writing Home* (1994) focuses upon his northern identity as a Leeds grammar school boy as did Richard Hoggart in *Uses of Literacy* when writing about the scholarship boy. For Hoggart problems of self-adjustment for the scholarship boy, whilst managing the transition from working-class to middle-class culture, are 'especially difficult for those working-class boys who are only moderately endowed, who have sufficient to separate them from the majority of their working-class contemporaries, but not to go much further ... but this kind of anxiety often seems most to afflict those in the working classes who have pulled one stage away from their original culture and yet have not the intellectual equipment which would then cause them to move on to join the 'declassed' professionals and experts. (Hoggart 1957: 293). The scholarship boy becomes more and more isolated from his gregarious working-class parents and peer group and becomes more self conscious, uprooted and isolated.

But Alan Bennett was no ordinary boy nor did he attend an ordinary school. Leeds Modern School was an aspiring state grammar, with a headmaster, as in The History Boys, anxious to be accepted into the independent/public school network of the Headmasters' and Headmistresses' Conference (HMC). He was no ordinary scholar either and, to the amazement of his peer group, appeared on the first day of school examinations in a suit. Such a penchant for formal attire heralded what was to come during Oxford matriculation. By his own account he 'cheated' his way into Exeter College, Oxford, and also during his undergraduate examinations, not by cramming and cribbing, but by turning historical questions on their heads tellingly, with a view to developing strikingly unconventional answers and 'getting noticed' by the examiner, much as the media historian Irvin counselled the boys in the play. He claims to be a failed academic, a failed Oxford historian to boot, on account of his failure to teach well and write up his research. A major factor too, for a medieval historian, was his failure of memory, a fatal problem for those who wish to practise empiricist historiography. But it is hard not to feel that this is false modesty – by his own dictum all modesty is false by its very nature. He draws on his historical training to write plays and filmscripts for The Madness of King George and also The History Boys. But the transition from Oxford medieval historian to national playwright is a reflection of his own celebrity. This is a different world from the ivory towers of the academy as the title of the original film was to be The Madness of King George III, but there were concerns for the American market that audiences would think this a sequel to Madness of King George I and King George II.

Finding a Voice and Losing One's Accent

His introduction to Talking Heads (1986) acknowledges that literary forms often dictate themselves and Alan Bennett justifies his innovative decision to write six half-hour monologues in the BBC TV series, Talking Heads. He argues that none of the narrators are telling the whole story in Talking Heads, whereas a more objective view would be to write in play form where we could meet all the characters and achieve a more rounded objective view. But this helps Bennett present the flawed characters and invites the audience to piece together what has happened below the surface of their seemingly ordinary lives. It is a world replete with dark sexual forces – suggestions of paedophilia, sexual interference and assault, pornography, innuendo and betrayal. In one sense Writing Home (1994), his annual January Diaries for London Review of Books, Untold Stories (2005) address these silences in the monologues by publishing his diaries, addressing topics which are taboo and naturally sit uncomfortably with his media designation as a 'national treasure'.

When writing plays Bennett usually deploys a metropolitan voice and form, whereas unsurprisingly his diaries are more northern and personal. What makes Bennett attractive are his asides, more accurately his insides, as he often makes public his personal thoughts, opinions and foibles. So when making a film about the changing nature of class for the BBC, located in the foyer of a Harrogate hotel, he comments on class, respectability and manners. In his diaries, he admits that the final product is mainly autobiographical, far more so than originally intended. Indeed, in the rare interviews, the favoured journalist never fails to comment on the presence of Bennett's Leeds accent which surfaces at times during their conversations. It is as if his Leeds childhood experiences are always subterranean, bubbling and breaking the surface of his Oxford and metropolitan demeanour, like streams in Yorkshire limestone country. Just as in his latest diaries for 2007, he mentions that he has favourite watering places between Arthington and Harewood to urinate when old age makes travelling by car more problematic for the bladder, but using the stately home of Harewood House as 'a place of worship' is beyond the pale.[8]

It took Alan Bennett quite some time to realise that his own childhood and adolescence growing up in Leeds was a worthy topic for his own writing.

> I don't think I saw Richard Hoggart as a cultural pathfinder out of Leeds. When I read him, c in 1963 I was already in New York and had left Leeds behind. What Hoggart did for me was to take me back to Leeds and to show me that there was a real childhood and youth there which I had gone through but which I'd never until that moment considered a viable – or feasible – background, something that I could write about. Had I read D.H. Lawrence at that point I might have come to the same conclusion by a different route but I hadn't-and also, probably, the passionate emotions of it would have put me off. Hoggart was more mundane and writing about the kind of streets where I'd been brought up – if not Armley, certainly in Wortley where my grandma lived.[9]

Lessons of History – Relevant and Useless Knowledge

There are strong resonances in The History Boys with Bennett both finding a voice and a vocation as suggested when he worried about cramming to get into Sydney Sussex, Cambridge or to secure an open scholarship at Exeter College, Oxford. There was an interesting conflation of reasons for finally settling on Oxford. He claims that part of his attraction to going for a scholarship at Oxford related to a crush on a fellow officer cadet, when learning Russian as part of his National Service, served in Cambridge in Russian and Language Services (source of Another

---

8   See Alan Bennett '*What I didn't do in 2007*', London review of Books, Vol. 30, No 1, p. 3.
9   Personal communication: letter 24.10.2006.

Country, Blunt and the rest). But he was also driven by vanity because he wanted to sport a scholar's gown. Ordinary undergraduate gowns were without sleeves so looked like riding jackets. But the quest for winning a scholarship must in major part relate back to his Leeds Modern School days and the premium placed on boys who could win scholarship places, rewarded by having their names recorded on Honours Boards in the main school hall. Indeed, successful Oxbridge boys would return to the school with 'insider' information, readily absorbed by the Headmaster as to what colleges were looking for in undergraduate recruitment and so act as 'hot tips' on which colleges could be easy interview targets in the forthcoming academic year.

The fear of 'cheating' in education reappeared when Bennett was studying for his finals. In an uneventful two years studying at university, he had approached his final examinations confident, as were college tutors for their part, that he would gain a solid second classification in his degree. He then suddenly recognised that success depended not so much on detailed reproduction of facts but rather on taking a unique and controversial line on conventional historical discourses. This tension is highlighted in the various classroom styles of the history teachers – Irwin (theatrical pyrotechnics), Hector (classical authorities, sheer love and the erotic nature of learning and education) and Mrs Lintot (historical empiricism). In the 1990s, interestingly enough, Oxford University explored the lack of women gaining firsts in history. An American professor (female) was drafted in to lead staff development – students were set mock exams which were blind marked by (male) internal markers and there were feedback session for markers to justify how they had awarded marks. Marks were awarded more for gladiatorial pyrotechnics of argument (a feature of male candidates and Oxford teaching style) whereas more industrious, empirically based, measured discourses (a characteristic of female students at Oxford) were marked down.

North, South and the 'Other'

Alan Bennett's celebrity is represented as one of coming from the North, of having an 'ear', of picking upon, valuing snippets of ordinary conversation, topics of ordinary everyday life and rendering these comments humorous and subversive. Most fans can loyally recount their own apocryphal vision of how he travels invisibly on the top seat of a Leeds tram, or more latterly a number one Leeds bus, overhearing and feeding on conservations which will eventually be transposed into one his plays or characters. But what startles about Bennett's work is his gentle amplification of silenced voices in working-class culture – the middle-aged bachelor looking after his mother, the fiercely independent elderly woman living alone, the resilience of people in a residential care home. They are all people on the receiving end of class inequalities, all slightly flawed, struggling to maintain dignity and respectability in the face of uncaring professionals, indolent organisations, dysfunctional and difficult families.

The 'Other' for Alan Bennett was not the postmodern ' Urban Other' documented in Manchester and Sheffield by Ian Taylor in his ESRC project to research 'the public sense of well-being: a taxonomy of public and space' (Taylor, Evans et al. 1996) nor is it the 'vagabond' theorised by Zygmunt Bauman as distinct from his two other categories of traveller – the pilgrim and the tourist (Bauman 1996). However, this would be familiar territory for Bennett who accepted a homeless and eccentric Miss Shepherd to settle for several years in his drive in Camden Town. As with much of Alan Bennett's output this bizarre tenuous relationship becomes a resource for his writing in his short story (Bennett 1990) which later in 1999 was produced as a play with two characters as Alan Bennett with Bennett acting one of his selves.

As a writer Bennett appears both detached and vulturine, constantly monitoring everyday life for possible material to be recycled into plays for theatre and television. His sense of 'Other' has been sustained by his equivocal sexuality – the early realisation that he felt somehow different from others. Such worries were clearly pressing during his childhood and adolescence, epitomised by his late puberty and arrested sexual development and experiences. Tensions resurfaced later on in his life when he had to choose between visiting and looking after his mother in the North, whilst starting out on his newly liberated sexual life in both London and New York. The rigid boundaries of his parents' respectable working-class culture failed to quell his prurience and he soon found himself surrounded by celebrities including David Hockney, Harold Pinter, Michael Frayn, Russell Harty, Michael Palin and the like. Bennett mentioned his parents' delight when meeting Russell Harty and the common interest and pleasure between Alan Bennett's father playing the violin to Harty's accompaniment on the piano.

> 'He's grand, is that Russell,' Dad said.
> 'We'll he's one of them,' I ought to have said, but if I had they probably wouldn't have minded. It was only if it got into the family that there was cause for concern.
> (Bennett 2005: 44)

Alan Bennett's Leeds in the 1940s and 1950s is essentially white. The major structuring processes are around class and education. Ethnicity is central to Bennett but it is the white ethnicity of Albion … dealt with in his plays such as The Old Country , 40 Years On, in his explorations of English churches, English landscapes in the Yorkshire Dales and naturally in An Englishman Abroad (with his recurrent themes of Cambridge and the monarchy, loyalty and deceit). In his later works, Bennett acknowledges race as a factor in people's lives but it is often through the eyes of white racism – with key voices complaining about 'blacks are taking over the country', or a temple built in the local crematorium, Asians 'not knowing the difference between what is normal and what isn't', a black home help, Zulema, not having the right approach to dusting and cleaning the council flat and so on. Another stereotype is of one of the local Asian shopkeeper but who, as a 26 year old hockey player, turns out to be a force for good for the alcoholic vicar's wife – not only providing comfortable sex at the back of the shop on a bed of lentils but also providing much needed alcohol on Sunday nights, as well as encouraging the woman to seek help from Alcoholics Anonymous. But the relationship is short lived with his move into an arranged marriage and consequent return to the Indian subcontinent. But the characters are framed through the eyes of the indigenous white population. Ethnic minority pupils are introduced in the The History Boys but Alan Bennett's own schooling was devoid of black ethnicity, although an undercurrent of anti-Semitism has never been far away in the history of Leeds schooling. Jewish North Leeds and anti-Semitism had a distinctive presence both in Leeds Modern School itself and in the imaginary of white working-class Leeds culture but it remained unrecognised or avoided. There are other silences in his diaries – particularly around sport – the holy trinity of Leeds United, Leeds Rugby League Club and Yorkshire County Cricket central to most Leeds' masculine and many feminine childhoods and identities rarely merit a mention.

So Alan Bennett remains both a quintessential Northern celebrity from Leeds and also a something of a paradox. He chooses to live in London and New York, whilst having a holiday cottage in the Yorkshire Dales; he is a fiercely private person yet his life has been exposed to media scrutiny. Nevertheless, he is very much a product of and a political defender of modernity whilst also embracing postmodern differences both in his writings and his lifestyle. The nostalgia for Leeds in the 1940s and 1950s with its respectable families and solidaristic local networks are all part of the postmodern condition. His generation was the harbinger of things to come and through humour, sexuality and media lifestyle, celebrated, subverted and undermined traditional working-class culture. Bennett himself functions as a bridge (that quintessentially post-modern millennium symbol) between modern and postmodern Leeds. His life and work span the gap between the city old and new, between the twentieth and the twenty-first centuries, between the back streets of Armley and the stage of the West Yorkshire Playhouse. Whatever hackneyed binary is conjured up to capture city changes – from steel bars to wine bars, from printing presses to garlic presses, from welding shops to Harvey Nichols, from slag heaps to financial and legal tips, from 'going out' to 'coming out' and so on, it is all home territory to Bennett's wit and humour. As an author he works as a post-modern chronicler of the modern and he is one of the best Leeds has produced.

References

Abrams, P. (1981). *Historical Sociology.* Shepton Mallet: Open Books.

Bauman, Z. (1992). *Intimations of Postmodernity.* London: Routledge.

Bauman, Z. (1996). From Pilgrim to Tourism – or a Short History of Identity in Hall, S. and du Gay, P. eds.

Bennett, A. (1988). *Talking Heads* London: BBC Books.

Bennett, A. (1990). *The Lady in the Van.* London: London Review of Books.

Bennett, A. (1994). *Writing Home*. London: Faber and Faber.

Bennett, A. (2005). *Untold Stories*. London: Faber and Faber/Profile Books.

Bramham, P. (2005). Habits of a Lifetime: Age, Generation and Lifestyle in Bramham, P. and Caudwell, J. eds.

Bramham, P. and Caudwell, J. eds. (2005). *Sport, Active Leisure and Youth Cultures*. Eastbourne: Leisure Studies Association Publications, No. 86.

Dennis, N., F. H., et al. (1969). *Coal is our Life: an Analysis of a Yorkshire Mining Community*. London: Tavistock Publications.

Hall, S. and du Gay, P. eds. (1996). *Questions of Cultural Identity*. London: Sage Publications.

Harrison, T. (1985). *V.* Tarsett, Northumberland: Bloodaxe Books.

Hoggart, R. (1957). *The Uses of Literacy*. Harmondsworth: Penguin Books.

Hutton, W. (1995).*The State We Are In*. London: Jonathon Cape.

Jackson, B. and Marsden, D. (1962). *Education and the Working Class*. London: Routledge and Kegan Paul.

Martin, B. (1981). *A Sociology of Contemporary Cultural Change*. Oxford: Basil Blackwell.

Pearson, G. (1976). Paki-bashing in a North-East Lancashire Cotton Town in *Working Class Youth Culture*, edited by G. Mungham and G. Pearson. London: Routledge and Kegan Paul.

Roberts, R. (1971). *The Classic Slum: Salford Life in the First Quarter of the Century*. Manchester: Manchester University Press.

Seabrook, J. (1982). *Working-Class Childhood*. London: Gollancz.

Taylor, I. et al. eds. 1996. *Tale of Two Cities*. London: Routledge.

Tunstall, J. (1963).*The Fishermen*. London: George, Allen and Unwin.

Urry, J. (1990).*The Tourist Gaze*. London: Sage.

Urry, J. (1995). *Consuming Places*. London: Routledge.

Williams, R. (1963). *The Culture and Society 1780-1950*. Harmondsworth: Penguin Books.

Internet-based references

Inside Bennett's Fridge, Daily Telegraph [website] Last Updated: 12:01am BST 30/10/2004 Available at www.telegraph.co.uk/comment/telegraph-view/3612508/Inside-Bennett's-fridge.html. [Accessed: 30.06.2008].

Owen, S. (2005), 'The Abuse of Literacy and the Feeling Heart: The Trials of Richard Hoggart', *The Cambridge Quarterly*, 32:2, 147-76. Available at http://camqtly.oxfordjournals.org. [Accessed 20.07.2008].

Chapter 6

Leeds and the Topographies of Race: In Six Scenes

Ben Carrington

Space is political and ideological. It is a product literally filled with ideologies.

Henri Lefebvre, *Reflections on the Politics of Space*

I recognise the place, I feel at home here, but I don't belong. I am of, and not of, this place.

Caryl Phillips, *A New World Order*

'Dirty nigger!' Or simply, 'Look, a Negro!'

Frantz Fanon, *Black Skin, White Masks*

Nationality: WOG.

Author: Unknown, *Leeds City Police charge sheet,* 23 February 1969

What, what nigger?

Author: Unknown, *Eltham, South London*, 22 April 1993

Do you want some, Paki?

Author: Unknown, *Leeds City Centre*, 12 January 2000

Northern Scenes: Hoggart and Hovis in Hyde Park

The look was intense. In fact, if truth be told, it was more of a stare than a look. Either way, I was definitely the one being looked at. I did my best to resist. I returned the gaze and smiled. The expression on her young, pale face was not hostile. Far from it. More intrigue. Puzzlement even. She continued to look. In her mind, something was clearly not right.

Chima and I were killing some time on this late summer's afternoon while we waited for a friend. For once, it was relatively warm. So we sat idly chatting on the kitchen steps of one of many 'back-to-back' terrace houses near Hyde Park where, seemingly, the city's entire university student population lived.

I remember the first time I walked around this part of Leeds that merges into Kirkstall to the east and Headingley a little farther north. I confess (though of course I never told anyone at the time) that on that very first visit I hummed the theme tune to Coronation Street as I walked the narrow, partly cobbled streets, with corner stores, seemingly owner-less dogs chasing their tails, and row upon row of bricked 'two-up, two-down' houses. A classic northern scene. I kept expecting Ken Loach to appear at any moment and shout, "Cut!" as dolly grips ran out to reset cameras for the next take and Ken Barlow received advice from the writers for the next scene. I really am in 'The North' I thought, as I wandered trying to find impossible-to-locate intertwining streets that would start with the same name but end with 'Terrace', 'Road', 'Avenue', 'Close', 'Lane' and many other variations, seemingly designed just to throw off the outsider.

As I became increasingly lost, though trying to look as if I still had my bearings, and as the roads got steeper the 'Corrie' soundtrack was replaced in my head by Antonín Dvořák's *New World Symphony*, more popularly known in Britain at least as the tune for the Hovis bread television advertisements. Cornets, flugelhorns, trombones and tubas playing out an imaginary northern soundtrack to where I would spend the next three years. Or at least what passed for 'the north' in my stilted southern imagination. Sure, Corrie was set in Manchester and this was Leeds, and as I would quickly learn, and as David Conn (2008) reminds us his essay 'The Beautiful North', Lancashire and Yorkshire should never be confused. But for me, at that moment in the mid-1990s, I was now in the heart of the north.

For someone who had grown up in south London during the 1970s and 80s, my distorted perceptions of Leeds were informed by the comically mythical and nearly universally white, vernacular portrayals of northern-ness that the television script writers of Corrie and the Hovis 'Boy on a Bike' ads had constructed. True, as a youngster, I had occasionally experienced the working-class delights of a week's summer holidaying in Great Yarmouth (I always preferred the Carrington family trips to Western-super-Mare mainly because they offered a chance to see Botham, Richards and Garner at Clarence Park) and once a long weekend in Windermere in the Lake District. But until I went to Loughborough University in the early 1990s, my experience of life 'north of the Watford Gap' was pretty limited and grounded in a south-London sensibility.

The nostalgic version of northern working-class culture that the 1970s Hovis ads depicted bore about as much connection to contemporary realities as the director Ridley Scott's other great filmic invention, 'Blade Runner'. It was a fake northerness that was sold and that always felt as if it belonged more to the late 19th rather than the late 20th century. In truth, the name 'Hovis' was coined not by some Lancashire baker but by a London student and the famous 'Boy on a Bike' ad was actually shot in Dorset. As Andrew O'Hagan (2006) noted in the *Daily Telegraph*, shortly after it had been voted Britain's best loved ad: 'The Hovis advert is probably the nation's favourite because it plays on romantic images of English endurance and pride – very Northern ones, too, the brass band, the cloth cap, the row of modest houses and the coveted bicycle – although the long street featured is actually Gold Hill in Shaftesbury in Dorset, about as far from Wigan Pier as it's possible to go without getting your feet wet'. So much for northern authenticity.

The KLF collective's early 1990s techno anthem 'It's Grim Up North' didn't help dispel southern prejudices. The song's video depicted a rain-soaked, black and white, postindustrial nightscape of car lights and anonymous motorways that Ridley Scott himself would have been proud of. Wherever it is was – presumably the M62? – it looked bleak. As if to emphasize the dreariness, the song's 'lyrics' are a series of northern towns that the hooded lead singer reads out in a distorted monotone voice:

- Bolton, Barnsley, Nelson, Colne
- Burnley, Bradford, Buxton, Crewe
- Warrington, Widnes, Wigan, Leeds
- Northwich, Nantwich, Knutsford, Hull
- Sale, Salford, Southport, Leigh
- Kirkby, Kearsley, Keighley, Maghull
- Harrogate, Huddersfield, Oldham, Lancs
- Grimsby, Glossop, Hebden Bridge.

That's just the first verse. Sounding more like a reading of a Saturday afternoon's rugby league results, KLF's typically sardonic tune, which on the video ends with the words 'The North Will Rise Again', was as much ironic mocking of southern perceptions of the 'grimness' of the north and its postindustrial collapse, as it was a send up of towns that lacked the phonetic kudos of their southern counterparts. As one *Guardian* journalist put it, 'It's Grim Up North, by a pseudonymous KLF, wrongfoots the listener. A deadpan catalogue of northern towns, recited over rainy-motorway techno, suddenly blossoms into a rendition of Blake's Jerusalem, as if arriving at some socialist rave utopia' (Lynskey 2007: 4).

Speaking of socialist utopias, it wasn't just kitsch pop culture reference points that had shaped my perceptions of the north but my reading of some the classics from Cultural Studies too. Despite their seminal brilliance in showing the 'self-making' and resistant qualities of the English working classes, these texts worked to conceal as much as they revealed about working-class life. E.P. Thompson's *The Making of the English Working Class*, written while Thompson was teaching at the University of Leeds and living in Yorkshire, is a case in point. Thompson acknowledges that the book was, in his words, 'coloured at times by West Riding sources' (1963: 13). But the 'colouring', alas, didn't extend to either marking the whiteness of working-class cultures nor acknowledging the small but significant black presence among those populations – in West Yorkshire at the time and throughout Britain before – nor in complicating his narrative of how the 'new' working-class cultures forged out of the conditions of industrialisation and latterly consumerism were themselves intimately connected to Britain's imperial adventures 'over there'.

We would have to wait for the likes of Catherine Hall, Anne McClintock, James Walvin and Peter Linebaugh (himself a student of Thompson) among others for the historical links between the metropolitan centre and the colonial exterior to be brought into conversation. Thus despite his deep appreciation of the making of culture from below, Thompson's framework could not account for nor take seriously questions of race, apart from the fleeting and tantalising glimpses that we get of a racialised English identity and the presence of these dark, unnamed interlopers. Thompson describes one aristocratic traveler to the Yorkshire Dales in 1792 as exhibiting the same level of 'alliterative hostility' towards the northern industrial working classes as 'that of the white racialist towards the coloured population today' (1963: 189) but we learn no more about what these 'racialist hostilities' (in 1960s Yorkshire) might be or where they may have come from.

The Hunslet hero Richard Hoggart similarly failed to see ways in which Britain's imperial adventures shaped the identities of working-classes at home as he wondered 'Who are the working-classes?', the famous rhetorical question that opens *The Uses of Literacy*. Hoggart's starkly stated belief that 'Working-class people are not, we know, particularly patriotic' (1957: 85) serves to close down the connections between imperial, racialised nationalism and class formation. The racial denizens, against and through whom Hoggart's northern working-class natives are formed, lurk in the shadows of the text, suddenly ghosting through scenes in often bizarre ways, yet without generating comment or analysis.

Hoggart informs the reader of the intimacy of the neighbourhood spaces where gossip ensures that everybody knows everybody else's business, often in great detail: 'that this young woman had her black child after the annual visit of the circus a few years ago' (1957: 43). Or the rhyming chants of children's games: "I like coffee, I like tea. I like sitting on a black man's knee' (1957: 49). Here we have the white working-class version of the immaculate conception. The black child miraculously 'produced' by the peripatetic black circus performer – both of whose life experiences we can only imagine – producing an early if somewhat novel form of white misapprehensions about black life: absent black fathers spawning, in the words of Waynetta Slob, working-class 'brown babies'. And the stimulant products of colonialism – tea and coffee – neatly conjoined with the novelty of the black man's body, servicing white pleasure and leisure, are, in Hoggart's hands, simply another alliterative chant, of no more analytical significance than 'eeny-meeny-miny-mo' and 'tinker-tailor-soldier-sailor'. For a commentator so attuned to the nuances of the words, smells and everyday mores of working-class life these present astonishing omissions.

Hoggart's disavowals become all the more striking in light of later accounts of life in Leeds during this period such as Kester Aspden's (2008) powerful dissection of the life and murder of David Oluwale at the hands of Leeds police officers in 1969 after Oluwale had suffered years of brutal, unprovoked attacks by the boys in blue. A story that for many decades was shamefully ignored by the city's historians and biographers and is only now, as Caryl Phillips' (2007) moving portrait shows, recognised as an important episode in both Leeds and Britain's collective story.

Contra Hoggart, Aspden (2008: 36) paints a more disturbing picture of how life in 1950s Leeds was riddled with racial tensions that were barely concealed beneath the surface from any observer who wanted to see them:

The streets could be dangerous. A black man out on his own at night was a target for the thugs and for the police. A white woman with a black man was looked on as soiled goods. The wise couples learned to keep away from well-lit roads, slipping down the ginnels, as though they were breaking apartheid laws. 'Oh, it was shocking after the war', one Leeds woman told the *Yorkshire Post* in 1955. 'One man in Leeds came up to me and spat in my face. He said it was shameful for English girls to go about with niggers'.

Little wonder the travelling-circus-baby-fathers never stayed around. Paul Gilroy (1996: 236) charges that 'it seems impossible to deny that Hoggart's comprehensive exclusion of 'race' from his discussion of postwar class and culture represented clear political choices'. Intended or otherwise, Thompson/Hoggart accounts of working-class cultures produced a de-racialised (northern) working-class subject within which white racism disappeared.

However, after my move north my knowledge and appreciation of Leeds and the surrounding towns and cities would deepen, extending beyond what I had read, heard and seen. A brief spell playing semi-professional football for Guiseley in the Northern Premier league allowed me to sample the delights of Harry Ramsden's over-priced fish and chips, and a couple of years at Farsley Celtic where I discovered, well, Farsley. Travelling to play northern league 'giants' such as Whitley Bay, Stalybridge Celtic and Accrington Stanley for mid-week away games in the middle of winter gave me a more intimate knowledge of the landscape and character(s) of such places. A few successful seasons playing Sunday league football for the Fforde Greene (a pub on the border between Harehills and Chapeltown), including a final played at Leeds United's Elland Road ground – the pub has subsequently been shut down and turned into a grocery store after one police raid too many - and my adventures with Chapeltown's Caribbean Cricket Club, meant that I eventually could claim some kind of temporary, honorary Leeds 'Loiner' status.

But when I first arrived, 'Leeds' evoked a mixture of disaggregated sporting, cultural and geographical signifiers: the all-white strip of Leeds United Football Club and a football team that used to be good in the distant past (1992 excepted); cricket, Headingley, Boycott and Illingworth; woollen mills and factories; the Yorkshire Ripper ... and Chapeltown.

And so Chima and I found ourselves chatting one summer's afternoon, neither in Ridley Scott's mythical northern land nor the lily-white world of the *Coronation Street* script writers, nor even the de-racialised communities described by Richard Hoggart, but in the much more complex space of 1990s post-industrial Leeds. Home to one of the largest legal and financial sectors outside of London, the location of the new Harvey Nicks and one of England's liveliest dance music club scenes. 'Back to Basics' displacing the back-to-backs. Where the city centre's Victorian civic buildings stand alongside the service industries' (post)modernist structures of steel and glass. The 'new Leeds' as Phillips (2005) calls it, of tradition and modernity buttressed against each other. And still waiting for our friend.

As we sat there a small group of white kids no more than 8 or 9 years old were playing 'hide and seek', running occasionally from house to house as they made their way down the street. One young girl was crouching in our backyard. I pretended not to notice so as to not give away her hiding place. Momentarily she left her play world and began to stare incongruously at us. I tried to stare back.

After a few more seconds had passed she politely asked, 'Are you students?'. I was surprised – it seemed an odd question – and before answering tried to guess why she wanted to know who we were. Having previously lived in a so-called 'student town' I was acutely aware of very real tensions that are caused when 'indigenous' populations see their communities radically transformed by ever increasing numbers of, not always socially aware, students. The relationship often labelled as 'town and gown' or 'student blight' somewhat obscures the underlying class tensions between working-class locality and middle-class transience.

By then I'd been studying for my doctorate for a couple of years, armed with my newly acquired street-savvy local knowledge, or at least the fact that I could now find my way around Hyde Park without getting hopelessly lost, combined with my growing sociological literacy, I set about reading the girl and her somewhat strange question. This would be easy. I guessed that she had heard her father or maybe her mother voicing disparaging remarks about 'those bloody students' at home. They lived locally. They were a family (let's call them the Hoggarts) of white working-class Hyde Parkers whose rhythm of life had been radically disrupted by late night parties, traffic cones chucked into their gardens and cars packing the roads where once they could easily park. Maybe her dad, in his youth, had ridden his bike up the hill from Kirkstall before all these cars appeared, with a warm brown loaf under his arm, desperate to get back to see the latest happenings at the Rovers Return. Having stumbled into our presence, she was perhaps eager to find out whether or not we were 'one of them'.

'Does your dad not like students?', I asked, confident in my little piece of social analysis. 'No, he quite likes them I think', she replied. She asked her question again, becoming slightly impatient at my avoidance of her question. 'Yes, we're students', I finally informed her. She paused, frowned and replied without the slightest trace of malice in her voice, 'But you're black'.

Chima and I were totally thrown. We half laughed and half looked at each other in disbelief. We set about explaining to her that being 'black' and being 'a student' were not mutually exclusive categories (we didn't phrase it like that) and that there were indeed many black people who were university students, both in Leeds and elsewhere. Our rather feeble attempts at an anti-racist, pedagogic intervention proved ineffective as she was confronted with, in her eyes, a living, breathing oxymoron, a *black student* – in fact *two* of them. The confused girl's friends soon caught up with her. As she walked away she informed them, in an almost 'You won't believe this!' manner, 'They're students!', before laughing and running off back into the ludic world.

Chima, a black, male British postgraduate law student of Nigerian descent, and I, a second year PhD sociology student whose father also happened to be from

Nigeria, had been stripped of our student identities by a small white girl playing hide and seek. Our blackness had condemned us from the moment she saw us. There was no return gaze in that moment.

Clearly, in the area that that young girl lived the overwhelming majority of the university students she encountered, including those from out of town, were likely to be white. Especially so in the 1990s. It may well have been that we obviously 'looked' like students, perhaps due to our age, clothes and idleness. We clearly weren't working. We were acting, somehow, *like* students. But we did not fit her preconception of what a student must *racially* look like. The category of 'student' was marked as white. There was little variation. Or at the very least that particular subjectivity, in that particular location, was marked, at best, as *non-black*. The signifier and signified could not be sutured with the misplaced referents in front of her.

In that moment both space itself and the categories of legitimate presence or belonging to that space, were produced through a deeply racialised lens. What I call a topography of race was in play. Topographies of race mark and define space itself shaping how different places come to be inscribed with racialised regimes of power that get enacted against and through particular bodies. Our bodies 'stood out' in that moment by their undeniable non-whiteness. As Sara Ahmed notes, bodies stand out 'when they are out of place. Such standing reconfirms the whiteness of the space. Whiteness is an effect of what coheres rather than the origin of coherence' (2006: 135-36).

Leeds, as with most major cities, has a particular topography of race that manifests itself in complex ways. Or more precisely a series of topographies of race. Thus, this particular part of the city of Leeds, 'Hyde Park', is registered in the local imaginary as a 'student area'. Given the under-representation of black people, and particularly black males, within higher education more generally, 'black' and 'student' get produced as not only separate but in some sense antagonistic categories.

At another, deeper, level such fictive representations both reflect and reinforce wider perceptions that seek to position black people as intellectually inferior (to whites), as not quite being up to the intellectual rigours that are required of, say, reading for a degree ... or in fact reading at all. These long held forms of European colonial racism about the black Other, what Frantz Fanon famously names as the 'racial epidermal schema' (1986, 112), did not disappear with the collapse of old colonial structures and the debunking of 19th century racial science. Black cognitive inferiority remains a deeply ingrained and structuring feature of common-sense understandings of racial difference that is still reproduced within many parts of the Western academy (Younge 1996).

Leeds University, it should be remembered, was where Dr Frank Ellis taught Russian and Slavonic Studies for many years while arguing as an 'unrepentant Powellite' that Asians and blacks should be repatriated, that multiculturalism (along with feminism) was corroding Britain as it was based on the 'lie' of human equality, that the British National Party was the only party that had the right ideas for Britain although it was a bit too 'socialist' for his liking, and of course, that old favourite, that blacks are inherently intellectually inferior to whites (Halpin 2006, Taylor 2006). Ellis was merely repeating the racist ideologies of other academics concerning the supposed black intellectual deficit that Chris Brand – the then Edinburgh University psychologist who shot to infamy in 1996 by making similar claims but adding an argument for the positive effects of paedophilia into the mix – had made before and that were rehashed a decade later by the ramblings of James Watson who told *The Times* that the notion of intellectual equality is a myth: 'people who have to deal with black employees find this not true' (Hunt-Grubbe 2007).

'Are you students?' It is in these moments of exchange, of physical and psychic contact, that Blackness and Whiteness are created and defined as seemingly immutable categories. But, as Fanon argued many years ago, the 'facts of blackness', or more accurately the lived experiences of being black, are never a 'simple' matter of observation. Such encounters carry with them a complex history of racialised meanings and power. Fanon's recollection of being defined and marked by his blackness by a white French child, who sees him on the streets and who, frightened, starts to cry in fear, is a pertinent reminder of this process. 'Sealed into that crushing objecthood' (1986: 109), Fanon is burned and marked by his colour, unable to move outside of the discursive limits proscribed by his blackness.

But our Leeds experience was somewhat different to that classic Fanonion moment. It was not our blackness *per se* that the young Leeds girl found disturbing. She was not frightened by us as the young child had been of Fanon. She did not cry in horror. Indeed she initiated the conversation with us. She was if anything *intrigued* if not quite fascinated by our blackness. It was however that the cultural ascriptions associated with being black in this particular area of Leeds simply did not include us being students as well. It was a moment of post/colonial racism. A double bind. Of intrigue and interest *and* objectification and racialised inscription. All at once.

This new fact of blackness is one where 'race', nation, gender, region, class and place powerfully collide. Where blackness itself is measured in relation to narrowly prescribed social roles such as 'athlete', 'criminal' or 'entertainer'. It is a form of conditional blackness, aligned to a delimited sense of Britishness that operates to exclude certain people from the imagined community. Or only allows them into the story as if they have just, somehow, arrived yesterday. The birth pangs of black Britishness breaking free from old colonial tropes just as new post/colonial narratives attempt to reinscribe the logic of white supremacy.

The latest Hovis ad, released in September 2008 to commemorate 122 years of the product and premiered, of course, during *Coronation Street*, serves to keep this distorted history alive. The new ad features a young white lad running through a series of street scenes that episodically move through time as he turns each corner, the Hovis loaf safely tucked under his arm. While the cultural signifiers of the ad serve to highlight the contested class and gendered history of Britain – the suffragettes and the miners strike are referenced – the narrative is suffused with the nationalist pomp and monarchist ceremony of Union flags, First World War heroism and Second World War Churchillian stoicism. That tired chant of two world wars and one World Cup revitalised one more time to breathe life into the old bulldog spirit.

Save for a faceless black man or two walking away from the shot and an elderly 'Asian couple' wandering down the street when we presumably reach the '1970s', the only significant colour in this narrative of Britishness turns out to be the little brown loaf itself. Even if the bread itself isn't whitened, the 122 years of 'British history' that Hovis reinvents serves to whitewash colour out of the nation's story: 'As white today as it's always been'.

During a time of global financial crisis, when capital markets have imploded, banks are part-nationalised in order to save them from collapse, where average British incomes have stagnated, house prices are declining and unemployment rising, such representations offer a retreat into the simplicity of white nostalgia as a refuge from the uncertainty of the cosmopolitan, threatening present.

'Are you students?' Was there a shift in that moment? The racial fabric tearing and being reconfigured? Was it an affirmation – 'they're students!' – in which the possibility of blackness opened up to include the black cognitive thinking self? Or was it a moment of dismissal, a denial and refutation of the co-joining – '(they think) they're students!'?

The reproduction of global white supremacy and the complex performance of blackness in response are always manifest locally and regionally. They require that we map the articulation of race not just historically across time or analytically through our conceptualising but concretely as it registers and resides in space itself. And how racism tries to push certain bodies out of place. As Ahmed argues, racism can be considered 'as an ongoing and unfinished history, which orientates bodies in specific directions, affecting how they "take up" space' (2006: 111).

Or to put it another way, can we re-imagine what Leeds 'is' through an account of its geographies and often disavowed histories of race? To trace race through space and place. To avoid the metatheoretical temptation to extract race from the everyday practices of Leeds' residents and visitors, its tourists and locals, the indigenous and the transient, and instead to think Leeds horizontally, from street to street. To be able to say something about the possibilities of living with difference in Britain, of the intersections of class and race, of sexuality and gender, and how a city like Leeds is experienced differentially by different bodies in the same places. To hold on to an autobiographical account of moving from the south to the north, from London to Leeds, that attempts to avoid the solipsism of the personal and instead tries to read the structural critically through moments or *scenes* of embodied (dis)placement.

Such a task might return us to a more nuanced account of the city – and by extension of the region and perhaps the nation – that avoids invoking hackneyed city myths as authentic northern truths but instead offers a guide, a route, a map of sorts, as to how the contradictions of the post/colonial city that is Leeds are lived, breathed and embodied.

Setting the Scene: Invisible Borders

As the white security guard walked towards me I noticed that his uniform was slightly ill-fitting. He was probably in his late fifties and looked like he had 'seen some action' in his time though he didn't really seem agile enough anymore to deal with a real security problem should one arise. More information guide than hired heavy. His face exuded an officious yet friendly look. 'You need to take the stairs, first floor, turn right and the room is on your left', he said by way of an introduction. Slightly startled at the unsolicited directions, I thanked him and made my way towards the stairs.

I had been standing in the entrance of Leeds Civic Hall, located in the city centre in what is now called Millennium Square. Not for the first time during my early days in Leeds I must have looked slightly lost. I had arranged a meeting with Ron. He was the chair of the Caribbean Cricket Club (CCC), the team I would study for my PhD on sport, race and identity. By coincidence Ron also worked for Leeds City Council and was on the initial steering group (that also comprised the then Commission for Racial Equality and Leeds Metropolitan University) that was overseeing a research project into racism in professional rugby league that I, Jonathan Long, and a few other graduate students were conducting (Long et al. 1995). When I told Ron about my research interests he immediately said 'You have to study us!'

Ron was, as the ethnographic textbooks put it, my 'gate keeper' into the research field. Over the years Ron would prove a useful source of information about Leeds, politics in the city, the struggles of the local black community and the role of what turned out to be Britain's oldest black cricket club in all of this.

As I made my way up the stairs I thought it touching of Ron to have left my description and directions to his office with the security guard at the front desk. Northern hospitality I guessed. I clambered the final steps, turned right and entered the first door on the left. The spacious room was packed with people going from table to table signing pieces of paper and picking up leaflets. I soon realized that this wasn't in fact Ron's office but a reception area that was used for a Department of Social Security (as it was then called) event helping 'job seekers' seek jobs and giving advice for welfare claimants. I suppose I must have looked like somebody in need of some DSS assistance.

I eventually found Ron's office a couple of floors up. He had obviously left no such instructions with security. I told him where I had been sent. He thought it funnier than me. 'Welcome to Leeds'. Ron informed me I'd better get used to such moments. The everyday racial politics of the city would veer between the racially banal – the banality of racism – that was sometimes so absurd as to be comedic, to other 'moments' that would be more menacing in tone and effect. But the cricket club would provide a different set of issues as to what living in Leeds meant for its black residents. I arranged to meet Ron the following week at the club where he would introduce me to the players and the supporters. 'Can you play?' Ron asked. 'Yeah, sure'. 'Good, we could do with some new players. Bring your bat!'

It is an oftentimes overlooked fact that C.L.R. James's *Beyond a Boundary* evokes a spatial metaphor. James's text and the title of what many regard as his finest book pulls into tension questions of racism and racial identity, of politics and sporting contestation, of space and embodied ideological struggle and takes them straight onto the cricket field itself.

All of these themes are simply, but brilliantly, captured in his reworking of Rudyard Kipling's meditation on Englishness. 'What do they know of cricket who only cricket know?' In James's hands 'England' is both replaced and displaced by 'cricket'. The moment of metonymic exchange serves to rework Englishness itself. England becomes cricket becomes politics. The boundary is dissolved and deconstructed precisely as it is recognised. Space is stretched and traversed. But boundaries, despite their physical manifestations, are also imagined. Boundaries, as Anthony Cohen noted many years ago, are produced in the mind of their beholders, as well as on maps (1985, 12). They work not just to mark the line between us and them, always essential work in the symbolic construction of community of whatever hue, but also to constitute the community in the first place. Hence the existential threat that 'border crossers' are supposed to present to the community itself. If our borders are porous then who are we?

Chapeltown itself is generally perceived to be located within the spatial boundaries marked out by Scott Hall Road to the West, Roundhay Road to the East and Potternewton and Harehills Lanes to the north, although its exact cartographic location, as the sociologist and part-time Chapeltown biographer Max Farrar notes, is a matter of some contention. Farrar points out that on some maps of Leeds Chapeltown is not even present:

> As Ralph Ellison (1952) famously demonstrated black people are rendered invisible by whites. The smaller-scale Leeds' maps erase Chapeltown altogether. But while the cartographers make Chapeltown invisible, the myth-makers insist on representing the black residents of the territory everyone in Leeds knows as Chapeltown. This 'visible invisibility' – the contrast between popular vision and cartographic blindness – is maintained by the mix of social motivations which come into play as soon as popularly named Chapeltown becomes the topic of discursive manoeuvre, with the various representational categories reflecting the values and ideologies of 'race' and sex held by their proponents. (1996: 105)

If the cartographic identity of Chapeltown is officially disputed, its popular signification is not. Due to decades of local and national media (mis)representations (Farrar 1997, Aspden 2008), Chapeltown has come to be known as an area of criminal deviancy and sexual promiscuity. As one demographic profile of the area conducted in the mid-1990s starkly put it, 'The predominant media depiction of Chapeltown is negative. Chapeltown suffers from a bad reputation. The word itself is associated with crime, drugs, prostitution, disorder and danger. This is combined with its image as a place where black people live' (PRI 1996: 12). I soon discovered that after early evening many cab firms would suddenly have all of their cars booked for the night whenever I would call for a cab to take me from Hyde Park to Chapeltown.

It is important to remember that, demographically speaking, the perception of Chapeltown as a place where the majority of residents are black is factually incorrect. As Gilroy notes, the 'blackest areas of the inner city are, for example, between 30 and 50 per cent white' (1993: 34), and this is true for Chapeltown. Despite being commonly seen as a 'black area', census data from the 1990s revealed that just over 39 per cent of the population of Chapeltown is actually 'white', with a total black (that is African or African-Caribbean) population of around 28 per cent and a South Asian population of just over 26 per cent (Farrar 2002: 23). As Farrar concludes,

> Chapeltown's reputation, in some quarters, as the "black ghetto", derives simply from its spatial congregation of black people, and its misrepresentation in the public imagination, not from any inherent features of black people (2002: 24). There goes the neighbourhood.

Despite the difficulties faced by the city's Asian and black populations, the area, which has above average levels of unemployment and poor housing conditions (PRI 1996, Farrar 1997, Farrar 2002: 27) has a large number of community institutions that were established as mechanisms in the painful transition from migrant, settler communities towards full citizenship and belonging. In his historical documentation of the struggles of Leeds' black populations, Farrar (1996: 49) notes,

> For such a small population, the number of social, religious and political organisations which had been established is remarkable... It is these voluntary organisations, whose activities range from the devotional (mosques, temples, churches), the cultural (carnival, theatre, music), educational (self-organised schools, youth activities) to the political (lobbing, campaigning), which are the backbones of the black communities in Leeds and which reflect the resilience, energy and achievements of the communities.

One disturbing trend has been the gradual retreat of the commercial sector, or at least its transformation, which has had marked effects on the social life of Chapeltown. Thus, while the voluntary and to an extent public sectors have attempted to sustain civic life in the area, the mainstream private sector has withdrawn over the years. Whilst a number of essential services such as small grocery shops (due to competition from near-by supermarkets), banks (due to alleged fears about crime) and even the petrol station, have left the area, there has been a concomitant rise in the number of solicitors, housing associations and book-makers.

During the time I lived in Leeds the *Yorkshire Evening Post* reported that one of the last remaining banks in the area was to close due to 'fears of crime': 'CRIME FEAR SHUTS BANK: Community anger at blow to suburb' (Maguire

1997: 1). This despite local protests and even police testimony that there was little threat of crime to the bank or to its employees and customers. As I have argued elsewhere (Carrington 1998), this has meant that despite, and probably because of, this economic stagnation, the local political, civic and cultural organisations, of which the cricket club is a central part, have maintained, and even increased, their importance to the collective well-being of the area. Everyday forms of sociability, of community building and identity formation become increasingly vital ways of coping and surviving postindustrial decline and inner-city disinvestment. Each institution that remains becomes a testament to survival.

Caribbean Cricket Club's ground called 'The Scott Hall Oval' is located high up on Scott Hall Road, a busy A-road that runs north of the city. The high altitude of its location means the ground overlooks the city and for many of the club members it is a marker of the black community's struggles for recognition by the city. The plaque inside the modestly sized clubhouse proudly and simply confirming the opening as 'The Realisation of a Dream'.

Crowd Scenes: Sporting Antiphony at Farsley

The boundary at Scott Hall Oval never seemed to offer much by way of a separation of the players from the watchers. C.L.R. James notes, in tones reflective of CCC supporters, that the well known Shannon Club of Trinidad 'were supported by the crowd with a jealous enthusiasm which even then showed the social passions which were using cricket as a medium of expression' (1963: 54). The 'crowd' is thus an integral part of Caribbean cricket culture. Richard Burton links the *active* watching of cricket to aspects of the carnivalesque and to Caribbean street culture more generally, 'so that cricket, carnival and the street corner become overlapping expressions of a single underlying social, cultural and psychological complex' (1991: 9).

In this formulation Caribbean cricket is seen to reflect broader aspects of black cultural expression through which new forms of identity are created. Using music as his exemplar, Gilroy has suggested that the lines between self and other within the black arts movements 'are blurred and special forms of pleasure are created as a result of the meetings and conversations that are established between one fractured, incomplete, and unfinished racial self and others. Antiphony is the structure that hosts these essential encounters' (1993: 79).

Extending these observations to the sporting arena, we might argue that a collective, democratic and performative sense of racial self is generated from these sporting moments of call-and-response, thereby giving a constitutive role to the crowd itself in the construction of *collective* meaning making and identity formation. It is, according to Burton, this dialogic aspect of the crowd which gives Caribbean cricket its distinctive character. Burton (1991: 9, emphasis added) continues:

What gives West Indian cricket its unique creole character is, in a very real sense, just as much the participants as the players themselves, so that the frontier between players and spectators – the boundary-rope which, in England until a few years ago represented a quasi-sacred limit that no spectator would dare transgress – is, in the West Indies and *in matches in this country in which West Indians are involved*, continually being breached by members of the crowd to field the ball, to congratulate successful batsmen and bowlers and, in not a few instances, to express their disgust at umpires' decisions, the tactics of the opposition, and so on.

Dramaturgical aspects of Caribbean cricket culture that James and Burton allude to structured the sporting spaces of CCC. The intensity of such expressions varied from game to game and across the seasons, and it would be an exaggeration to suggest that there was always a 'carnivalesque' atmosphere at the club. Nevertheless, the general *habitus* of the players and spectators, particularly amongst the senior players and older supporters, did reflect a distinctive black pan-Caribbean performativity regardless of where in the Caribbean each player or supporter had come from.

CCC functioned as a sort of regional 'home from home' for all of the 'overseas' Caribbean players, that is those paid to play for various league teams, that played in and around Leeds. CCC was one place, a particular space, where cricketers from the Caribbean could come when not playing for their sponsoring teams where they could enjoy aspects of Caribbean culture. They would sometimes play for CCC in friendly games if their own team was not playing. As Harold – a brilliant batsman who had played in what was then the Red Stripe Cup for the Leeward Islands – pointed out, CCC should be 'projected to the overseas players who come here from Barbados, Trinidad, Guyana, Jamaica, that here is a home for anyone out there, "Come here because it's home". Because it's a different culture. Most of us who come and play in Bradford, at these white teams, it's different. Of course you would all enjoy cricket regardless of where it's played, whether it's the Caribbean or Farsley in the Bradford League, but there is a time in which you can enjoy your own culture when you come to Caribbean'.

The spectators' support for the players – and sometimes the criticism of them too – was constant and vocal. Car horns would be blown whenever a six or four was struck or when an opposition batsman was given out. Sound systems would sometimes be in operation during the weekend games if the clubhouse was later going to be transformed into a dance space, as it often was. On the warmer days when there was a large crowd watching the games, many of the supporters would offer 'advice' to bowlers, batsmen and more regularly the umpires, from the boundary edge – and occasionally inside the boundary too.

Burton argues that the expressive behaviour of joking, ridicule, and argument reflects a masculinised Caribbean street culture that is carried over into the cricket arena such that the 'genteel' aspects of English cricket, associated with the values of restraint, fair-play, gentlemanly conduct, and other classed and gender-inflected norms of behaviour, are reversed. Burton identifies 'the marked stylization and hyperbolization of verbal insult' which manifest themselves in 'expansiveness, camaraderie, unruliness, jesting, joking, verbal and bodily bravado, clowning, in a word playing' (1991:17).

The playing of dominoes – often aggressively, always loudly – was one aspect of this type of stylisation of self that spectators, and players too when not directly involved in the game, would often engage in. These aspects of CCC would often evoke surprised and worried looks from the supporters of opposition teams when they witnessed such verbal encounters. Again, Burton (1997: 159) argues that such exchanges seem to the outsider 'perpetually poised on the brink of explosion as ego grates against ego and the potlatch of boast and counterboast, insult and retort, mounts to a threshold beyond which violence surely must erupt'. 'But', notes Burton, this is largely a performance for

> ... that threshold is never reached, or if it is the whole ritual of reputation collapses with its crossing ... this is a world of stylized, not actual, aggression, and while on the surface the men are affirming, or attempting to affirm, their superiority *as individuals*, deep down what they are acting out in the form of a competitive verbal ritual is their equality *as a group*. (ibid.)

The effect of these expressive forms was to mark these couple of acres at the Scott Hall Oval as a *black space*. Unmistakably so. Outsiders could come in. And were often welcome. A chance to eat some curried goat and rice or maybe ackee and salt fish rather than cucumber sandwiches and scotch eggs. But this was one corner of Leeds where the white gaze could not travel easily. Where the everyday injustices of white racism could be temporarily suspended at the boundary edge, even as those discourses framed and flamed the intensity of the sporting battles.

Whilst Burton's anthropological eye often comes close to essentialising aspects of Caribbean male culture and ignoring the fact that such modes of behaviour are not unique to black expressive behaviour – these could be argued to be key features of working-class (male) cultures in general (Back 1994) – the performance of this culture, always inflected through a strong Caribbean patois that became stronger as the evenings and Dragon Stout consumption got longer, did give a distinctive feel to the club in contrast to the other (largely white) teams in the league.

That said, there was often a degree of contestation and consternation that would be aired by some members of the CCC committee about the suitability of such behaviour. Some of the players were part-time DJs and had a 'crew' called the Cockspur Crew (named after the Barbardian rum) who would offer the most vocal support. For example, Tony who was on the committee and the brother of Ron, tried to introduce a no-smoking rule during one season for those players and spectators who would sometimes smoke marijuana. Tony argued that aside from its illegality such behaviour was damaging the 'image' of the club and by extension Chapeltown as a whole.

For Tony and others, the club needed to appear 'respectable' to the league officials and other clubs. This was a tension between, in Burton's (1997: 158) terminology, the street culture system of *reputation* and the more conservative demands of remaining *respectable*. The no-smoking-in-public rule was observed for a few weeks, and it was collectively agreed that players should not smoke inside the changing rooms or during the game. In the end though the non-playing members of the Cockspur Crew and their friends re-asserted their 'right' to act as they had previously and little actually changed beyond people being more discreet about 'rolling up' in public.

Burton suggests that 'the qualities that West Indians most prize in their cricketers are essentially "street qualities": what counts is not the mere scoring of runs, but scoring with style, panache, flamboyance, an ostentatiously contemptuous defiance of the opposition. Similarly, fast bowling must not simply, be fast but look fast' (1991:17). The aesthetisation of the player's bodily movement, in which the crowd responds to the artistry of the shot, reflects James's attempt to argue for cricket's place as an art form in its own right due to the composition of body shapes and the poetic charge of cricket. As James rather grandly noted: 'We may some day be able to answer Tolstoy's exasperated and exasperating question: What is art? – but only when we learn to integrate our vision of Walcott on the back foot through the covers with the outstretched arm of the Olympic Apollo' (1963: 211).

CCC players were expected to play in a particular style *for the crowd*. Whenever a CCC batsman struck a cover drive some of the crowd would shout out 'Hold it, hold it!'. The player was encouraged to hold their body position for a few extra seconds as the ball flashed to the boundary, in order to 'milk' the applause and shouts from the crowd and to let the bowler know who was boss.

Also whenever a CCC batter was hit on the body cries of 'Don't rub it!' would be heard across the ground. This supposedly demonstrated both the toughness of the CCC batter in not showing any visible signs of pain – and therefore weakness – and also in this refusal to give the opposing bowler the satisfaction of knowing he had 'hurt' the batsman. Similarly any balls bowled short of a length would be met with cries of 'Hook 'im, hook 'im!'. Short balls were there to be hooked, not ducked out of. Harold would often take on the quickies with hooks and pulls of such ease and quality that he would somehow manage to intimidate any opposition bowlers foolish enough to bowl any short balls at him.

For many of the supporters the CCC represented the residents of Chapeltown, the black community in general and even the wider pan-Caribbean diaspora (see Carrington 1999). For some, the CCC was a surrogate for supporting the actual West Indies national team. As Pete, the third team wicket keeper, put it to me, 'As far as I'm concerned we're just an extension of the West Indies national team'. This sense of identification was strongest during the 1995 season when the West Indies toured England. During home matches, television and radio commentary on the Test series would often be relayed in the clubhouse and across the ground. The connection was in some ways a literal one. The West Indies' player Stuart Williams who toured that year and finished with a batting average of 46.00 (higher than senior players such Carl Hooper and Richie Richardson), had played for CCC two years previously and was a close friend of Harold and some of the other players and supporters, especially those from St. Kitts and Nevis.

During the first Test match of that series a few of the supporters and players had gone down to Headingley to watch the game and complained that the officious stewards had prevented them from having 'a good time'. Apparently the only way spectators were allowed to enjoy the games was by polite applause. Brett, the first team captain, complained that he wouldn't go to Headingley that year, adding that 'The Oval [in south London] is the only place now where black man still rules'.

Due to supposed fears over crowd behaviour the England and Wales Cricket Board (ECB) had introduced increasingly draconian measures concerning what could be brought into grounds which resulted in a ban on, for example, musical instruments as these were deemed potentially dangerous 'weapons'. The effect of these restrictions meant that, as many commentators noted during the 1995 Test series, the degree of support for the West Indies team seemed much less vocal and numerous than in previous tours. Of course during that first Headingley Test when the West Indies batsmen did score a four or six, the obligatory camera shot of celebrating West Indies supporters meant that a few of the CCC members made it onto the television screens – much to their delight when they returned to the clubhouse later in the evening.

By contrast, Raymond Illingworth, who was then both England manager and Chair of Selectors was often the focus for abuse and scorn whenever he came onto the television screen. Chris Searle has argued that for many years Yorkshire had a reputation as 'unwelcoming territory for black cricketers' (Searle 1990: 43), in part due to a number of high-profile incidents of racist abuse from supporters at Yorkshire cricket grounds and statements, widely seen as racist, by prominent members of Yorkshire Country Cricket Club (Searle 2001).

During the England versus Pakistan Test at Headingley in August 1996 the Western Terrace faithful subjected the touring side and any Asian spectators unfortunate enough to be within ear shot to a torrent of vicious racist abuse that was so bad as to attract national media attention and condemnatory comment (Searle 2001). Mike Marqusee (1994: 143-144) too notes that

> The roots of racism in Yorkshire cricket are set deep in the county's peculiar regional chauvinism, a chauvinism warped by years of cricket failure ...The powers that be at Yorkshire have for many decades preferred the spurious roots of racial and cultural identity to the living roots of the game as it is actually played in the locality ... It is, at its core, profoundly exclusive.

Illingworth was seen to personify the traditional prejudiced *habitus* of the Yorkshire county cricket establishment; the archetypal Yorkshireman. This perception within the club was further reinforced when Dermot Reeve revealed in his autobiography that during England's 1995-96 tour of South Africa Illingworth had referred to Devon Malcolm as a 'Nig-nog' (Carrington and McDonald 2001: 63). Fellow black English Test bowler David 'Syd' Lawrence told the *Daily Mirror* he'd have hit Illingworth if he had heard that himself, adding, 'Illy is a typical Yorkie. He seems to love living up to the image of a hard-nosed Yorkshireman. I don't think I've ever played in a game against Yorkshiremen where there hasn't been a touch of racism' (Lawrence 1996: 33).

Later, when Imran Khan raised the question as to why Yorkshire County Cricket Club had failed to produce any home-grown Asian players, Illingworth typically rejected any talk of racism in Yorkshire as 'ridiculous' (Illingworth 1999: 13). The club's well documented history of racial exclusion (Long et al 1997), that continued long after it dropped its requirement of having to be born in Yorkshire to play for the county and a key reason why so many Asian leagues were established in the first place in what Chris Searle once called a '"Jim Crow" cricket arrangement' (Searle 2001: 200) was deliberately ignored. Illingworth turned the argument around and accused Asians of selfishly rejecting YCCC's sincere attempts to reach out and instead suggested that it was the Asian community itself that had engaged, for some unspecified reason, in racial separatism: 'they generally seemed more comfortable staying in their own environments. That, to be honest, is the major factor behind the failure of [Asian] players to come through' (1999: 13). Also in April 1997 Illingworth opined to *The Sunday Times*:

> A lot of the Asian people are not strong enough to bowl quickly. How many of the Indians have ever bowled quick? Pakistan have had one or two but only in the past few years. They are generally from northern Pakistan where the tribesman are quite tall and strong. It is a fact that a lot of West Indians, because of their physique and their looseness, can usually bowl quicker than white people. (cited in Carrington and McDonald 2001: 63)

As a telling indication of how the county cricket team were viewed, although a few members of CCC supported Leeds United Football Club and would occasionally wear Leeds shirts, no one supported Yorkshire County Cricket Club. YCCC's continued resistance to fielding a British-born black player (Conn 2006) only served to reinforce the perception of YCCC as the old, white boys club. As Conn notes (Conn 2006: 8), 'On the coaching, development and administration side, Yorkshire CCC does not have a single black or Asian employee'.

Even LUFC, despite the notoriously racist elements of its fan base has a history of black African stars stretching through four decades from Albert Johansson to Tony Yeboah and Lucas Radebe. The writer and long time LUFC fan, Caryl Phillips captures this dualism of the *amour et haine* relationship that 1960s and 70s black Loiners have had to struggle with when he writes:

> The same people who would hug you when Leeds scored (which we often did), would also shout "nigger" and "coon" should the opposing team have the temerity to field a player of the darker hue …. The sight of fans in the Leeds Kop

swaying and singing "I'd rather be a nigger than a scouse" was a little difficult for me to compute. (2002: 299)

Yet despite Phillips' traumatic relationship to LUFC it is one he comes back to time and again in his writings. The last piece in the section 'Britain' from Phillips' selected essay collection *A New World Order* is entitled 'Leeds United, Life and Me'. No such pained odes to Yorkshire County Cricket Club or its players exist within the extensive Phillips *oeuvre*.

In 1995 CCC made it to the semi-finals of the Evening League Trophy, a quick-fire game of 20 overs-a-side (now popular at county and international level as Twenty20 cricket) in which teams from the various Yorkshire leagues competed. No team from the Leeds League had ever made it to the final and a large convoy of players and spectators set off for the game against the Bradford league side Farsley.

The game was given heightened significance, as this was the home team of none other than Raymond Illingworth. It's not often you get a chance to stick it to The Man. But we had a chance. Black and white photographs of Illingworth's extensive county and national career adorned the walls of the Farsley clubhouse. Illingworth of course was also, at that time, the last England captain to have successfully won a Test series against the West Indies, way back in 1969. By chance Illingworth was at the game that night, hanging out by the clubhouse and occasionally walking around the boundary, and was constantly offered very public 'advice' throughout the evening by the CCC supporters on how he should manage the England team.

Farsley batted first in front of a large crowd, just about evenly split between Caribbean and Farsley supporters. The game was tense. It was the first time I could remember being nervous whilst fielding. I dreaded dropping an easy catch. Fortunately, Clive was his usual brilliant self, at one point running out a Farsley batsman with a direct throw from nearly 50-yards. Illingworth was apparently impressed and asked 'Who's that?' Tony told him and added that there was a further wealth of talent down at CCC if he wanted to check it out. Our bowlers, led by Brett, kept a tight line against some strong Farsley batters and we fielded as well as we could. Farsley eventually made a fairly decent 122 runs for 7 wickets – a run rate of just over 6 an over. They seemed confident of victory as their final two batters left the field to applause from their supporters.

We made a good start in reply. We were 56 off the first 8 overs with no loss of wickets and going well. But then a couple of wickets fell and the run rate slowed. The earlier enthusiasm of the CCC supporters waned slightly as the required runs per over increased to 8, then 10, then 12 runs. Harold continued to bat well but the Farsley players and supporters became more and more confident which each passing over. Come the last over we needed 16 runs for victory. Fortunately, Harold – our own Brian Lara – was still at the crease. It had come down to this moment of reckoning: Leeds League versus the Bradford League, Caribbean versus Farsley, Chapeltown's dreams resting in the hands of our best batter.

Farsley's 'overseas' player was a six-foot plus white Australian fast bowler. Their captain ensured that he was given the ball to bowl the final over. Six balls, 16 runs needed to win. With both sets of supporters and the England manager and Chair of Selectors looking on, the Aussie thundered the first ball down at Harold who leaned into the shot, and stroked it powerfully through the covers to the boundary for four. A few of the CCC supporters momentarily ran onto the pitch, somewhat prematurely, to celebrate. The sound of car horns suddenly erupted across the ground. Maybe we had a chance. Five balls left, 12 to win.

The Australian looked pissed. He came in again, faster this time and pitched the ball well short of a length. The ball rose sharply towards Harold's chest. Harold barely moved, stood his ground and pulled the ball for a huge six over the head of a despairing Farsley fielder. Caribbean spectators yelled out, more car horns were blown, the Farsley players looked on anxiously. From the brink of defeat (sixteen runs off the final over against one of the Bradford league's best bowlers?) it suddenly appeared that Harold might just pull it off. As long as he didn't get out. Six more needed off three balls. Game on.

The Aussie seemed to lengthen his run up clearly trying to intimidate Harold. He came in hard and pitched a fast, nasty bouncer. Harold, playing as if he was practising in front of the clubhouse with a tennis ball, simply rocked back onto his heels and struck the ball even further than the previous shot out of the ground for another huge six! The players and CCC supporters couldn't believe it. We'd done it. We all raced crazily onto the pitch and carried a slightly embarrassed Harold off the square on our shoulders. The Farsley players looked totally devastated by the loss. They weren't expecting to lose to the boys from Chapeltown. Not at home.

As we left the field Illingworth was goaded by our supporters. One of them shouted out to Illingworth, 'That's how you win a game Raymond!' Illingworth just smiled but remained silent. The previous week England, under Illingworth's stewardship, had lost the third Test to the West Indies by an innings and 64 runs within three days. England's second innings total of 89 was their lowest ever in a Test at Edgbaston.

True, our win was not quite the series 'Blackwash' that greeted England teams during the 1980s when they would face the might of Clive Lloyd, Viv Richards, Malcolm Marshall and Michael Holding. But for the players and supporters on that summer's evening, the location might as well have been Lords or Headingley and the victory had the sense of being 'our' one-day Test win. It was only a game. Only one night. But it felt good. It felt, as Pete had said, that we were an extension of the West Indies team and them of us. Sadly, despite the invitation, Illingworth never made it to the Scott Hall Oval to check out the young cricketing talent.

Public Scenes: West Yorkshire's Finest

'Where have you been tonight?' It was getting stopped for the third time that really pissed me off. The walk from Beckett Park to Hyde Park took about 40 or so minutes. Normally I'd try to catch the bus back down the Otley Road towards the City Centre. But depending on the time of day, the amount of traffic and how many books I had to carry, I'd often walk. While most of Hyde Park towards Headingley Stadium was studentville, the other side of Headingley going farther west was suburban and middle-class with semi-detached houses rather than the traditional back-to-backs. Maybe it was just bad timing. Maybe there had been a string of offences committed by another tall black guy with dreadlocks. But in my first year in Leeds while walking back home from campus I was stopped three times by the West Yorkshire police doing 'routine patrols' on the same stretch of road. A small addition to the nearly two million 'stop and accounts' carried out by the police, where black people are two and half times more likely than whites to be subject to such interrogations (Travis 2008).

The first time was a bit unnerving. A car suddenly pulling up alongside me. Officer coming out of his car, stopping me in the street and asking where I'd been, where I was going and what I was doing in the area. People passing by wondering what I must have done to solicit such attention. I even began to wonder what I might have done. Other than being black in Headingley at night.

The time didn't really seem to matter. Sometimes early evening, sometimes much later, either way I'd get stopped. The second time I thought, 'that was a bit of a coincidence'. I kept wondering if I'd see this 'other guy' whose description I evidently fitted. But the third time I was pissed. I asked the officer for his name and number. I think it was on his shoulder anyway. I wanted to start keeping a record though I knew it would be useless. I'd forgotten whatever information I had by the time I got home as my mind raced at the unprovoked harassment.

Drawing on Fanon, Ahmed notes that 'racism "stops" black bodies inhabiting space by extending through objects and others; the familiarity of "the white world", as a world we know implicitly, "disorients" black bodies such that they cease to know where to find things – reduced as they are to things among things' (2006: 111). It was this reduction that got to me after a while. But also the sense of powerlessness. Knowing that if I pushed back too hard that, well, that anything could happen. And without witnesses the super-human black body doesn't seem to be so strong once it reaches closed police cells. One officer left me with a 'Be careful!' warning as he got back into his car. But it wasn't clear who or what I needed to be careful of. Though I suspected it was actually of him.

A year later myself and two other black Farsley Celtic footballers were pulled over on our way back into Leeds for doing 38mph ... in a 35 zone. We knew our exact speed as the police car on the inside lane was driving painfully slow and Matt's white Astra GTE had a digital speedo. We passed the police car, watching the LCD display showing 38mph. West Yorkshire's finest decided that constituted a serious driving offence and his lights went on. After all of us were told to get out of the car and asked the usual perfunctory questions, Matt was 'let off' with a warning. I wanted to get the officer's details but Matt and Wayne laughed if off and told me to forget it. They'd seen worse and in their eyes at least the fool had been polite.

And once me, Chima and our girlfriends were standing outside Cottage Road Cinema waiting for a taxi to take us into the city centre. We'd been waiting no more than five minutes when from nowhere a car suddenly pulled up onto the kerb. I swung around to see two guys and a woman jumping out of the car. One flashed a badge in my face, the other one went towards Chima and the woman stepped towards our girlfriends. It quickly became apparent that we weren't being jumped by some random white thugs but by plain clothed West Yorkshire Police officers. They claimed there had been reports of drug dealing in the area and Chima and I, as ever, fitted the description. We were searched and had to account for where we'd been that evening and what we were doing 'loitering' on a side road. After a few more minutes of questions and fact checking the plain-clothed officers looked somewhat disappointed that the two six-foot black drug dealers they were about to bust turned out to be a doctoral student and the other a born again Christian more likely to be carrying a copy of the Bible than a bag of weed. Maybe they got back into the car saying 'They're students!'.

These were the everyday cuts of racism designed to wear you down and remind you as to where you should belong. Not here. Not in this part of Leeds so far from Chapeltown. It wasn't as if I was unfamiliar with these forms of cat-and-mouse, stop-and-search, meant to humiliate. London's Met Police were just as adept at such games. Just that this was new territory for me.

The 'strange' and the 'familiar' should not be read though as absolute markers of spatial and cognitive difference. As Ahmed notes, 'Even in a strange or unfamiliar environment we might find our way, given our familiarity with social form, with how the social is arranged' (2006: 7). Inhabiting space itself involves a 'dynamic negotiation' (ibid.) between the strange and familiar. I understood the 'social forms' of whiteness. Recognising the unfamiliar as a (potential) threat, understanding the risks that certain bodies have in public spaces, is part of an embodied practical knowledge that is necessary particularly when those public spaces are defined by a whiteness that threatens the ontological security of black and brown bodies.

But even familiar spaces can become dangerous at a moments notice. Stephen Lawrence and Dwayne Brooks, two fellow south Londoners from the same part of London where I grew up, Thamesmead, instinctively knew that on the night when they met five white lads in Eltham after missing the bus home. The reaction time to recognising the threat that often appears out of a clear evening's sky can mean the difference between living or not. Dwayne was lucky that night. Stephen wasn't. And neither was Rolan Adams nor Rohit Duggal before him. But the point, without wishing to over dramatise a situation as traumatic as that April night in 1993, is that bodies of colour in public spaces have to constantly watch for threats in ways that most white people rarely recognise.

The Leeds Met University student Sarfraz Najeib found this out after the then Leeds United player Jonathan Woodgate and his pals decided to kick him down Mill Hill. Leaving a message of a broken leg, nose and cheekbone on the body of the young Asian lad for daring to disrespect their masculine whiteness. Fellow

LUFC player Lee Bowyer was acquitted of causing grievous bodily harm (though he later paid Sarfraz and his brother £170,000 in an out-of-court settlement following a civil case brought against him) despite his previous record of violence and the judge's conclusion that Bowyer's interviews with the police were 'littered with lies'.

As the journalist Gary Younge noted, the attack on Sarfraz Najeib 'did not take place in a vacuum. It took place in a police district where the deputy chief constable conceded to the Lawrence inquiry that his force did "not have the full confidence of the minority ethnic communities". It happened in a city which has seen incidences of racial harassment increase almost 13-fold between 1995 and 2001; and where one of the sharpest rises has been in "attacks including actual/ grievous bodily harm"' (2001: 8). Being brown or black in any part of town on the wrong night at the wrong moment can leave scars or worse on those too slow to run or react.

Yet the hands of the law – as the bodies of David Oluwale, Joy Gardner, Brian Douglas and hundreds others known and unknown can testify – in the process of stopping, searching, interrogating and detaining can be as deadly as the knives and fists on the streets. The cumulative message from these assaults, be they symbolic or actual, is that these public spaces are white. The boundaries of identity whether imagined as geographical markers or the symbols and cultural signifiers of institutions are defined in these moments where the 'power geometry', as Doreen Massey (1993) puts it, of white space is produced and exercised on the bodies of Others.

Making a Scene: When Caryl met Raymond

When I first saw the list I actually thought it was a spoof, a slightly late April Fools joke. But the university's honours list turned out to be serious. One way in which academic institutions attempt to create narratives for themselves, to say something about the type of intellectual space that they promote and their place in the community, is by the award of honorary degrees. In April 1997 the management of Leeds Metropolitan University decided that Raymond Illingworth should receive such as degree because he (along with the other awardees) embodied through his achievements and work the 'values with which the University would wish to be identified'. Faced with such an affront Max Farrar, then teaching sociology at Leeds Met University, and I decided to let the Vice Chancellor know this wasn't a great idea.

For sure, in the grand scheme of things this was hardly going to be the greatest political protest that Leeds had seen. But given the discredited reputation Illingworth had at that point, particularly within the local Asian and black communities, some form of protest had to be made, however small. Sometimes symbolic gestures are all you have. C.L.R. James once noted, in conversation with Stuart Hall, that George Padmore would always agitate whenever something was occurring within the Empire that the Colonial Office would rather go unnoticed. Meetings would be called to discuss the issue at hand and regardless of how many people turned up, a resolution would be passed deploring the British Government's latest action and letters of protest delivered to the Colonial Office. Padmore's advice to James was clear:

> Never let anything happen without your doing something about it. Because if three or four things happen and you don't do anything, they will go further, and when people protest, they will say, 'These people don't care. Whole things have been going on, they never pay any attention.' And he never let anything pass. (Hall 1996: 25)

So we acted. With an Open Letter, drafted by Max and myself, co-signed by 40 plus members of staff, spelling out Illingworth's dubious comments and actions over the years from his support of Rebel Tours to South Africa in contravention of the Gleneagles Agreement to the accusations of his racist treatment of Devon Malcolm. We pointed out the irony in awarding an honorary degree to Illingworth at the same time as giving one to Caryl Phillips. We noted that we hoped Illingworth would refrain from calling Phillips a nig-nog when they met. The letter was sent to the university's newsletter, inspiringly called 'The News', to be published before the ceremony in order to push the university to respond publicly.

Max and I were then invited to meet with the Vice Chancellor and Deputy VC. They said that they understood our concerns but made it clear that Illingworth had done lots of good work and they were not going to withdraw the award. They agreed to issue a reply to our letter in 'The News' re-stating the university's commitment to diversity and equal ops and the like but insisting that Illingworth was really a good chap who'd just been misunderstood by his detractors. The awards must go on.

When the next issue of the 'The News' came out our letter had not been published. Someone had decided that it would be in 'everyone's interest' if the story died down. So we responded by doing the next best thing and alerted the regional and national media to the protest. There were headlines in the *Leeds Student* newspaper: 'Bosses Honour Cricket "Racist": Award for "Bigot" Illingworth'; June's edition of the *Campaign Against Racism and Fascism*: 'University in "racist" award fight', and *The Guardian*: 'Illingworth in row over degree award'. It was also picked up on BBC TVs 'Look North' programme, the main regional radio outlets as well as on BBC Radio 1, 4 and 5.

The University hierarchy wasn't happy. Max was officially informed that this protest would damage his career. Threats of disciplinary action were issued to anyone who might be thinking of mounting a demonstration at the ceremony itself. Any disruption would be deemed as 'bringing the university into disrepute'. A somewhat ironic charge as this was precisely what the university itself had done by deciding to honour Illingworth. But one person's disrepute is another's honorary degree.

The award ceremony was held at Leeds Civic Hall. I managed to find the correct room this time, without the help of the security guard. Leslie Wagner, a deeply unpopular VC among both the students and staff and who at one point hired security protection to prevent sit-ins from taking place in his office, sat on stage looking pensive and hoping that the assembled television crews and reporters wouldn't have a dramatic protest to show their evening audience. Illingworth got up to make his short acceptance speech which was painfully bad. He cracked a rather sexist joke about women not understanding sports vernacular, said a few obligatory thank yous, and then addressed what he considered to be the falsehoods that had been told about him in the press. And then he told the audience that he had black friends. A lot of them. And then he listed some of them. As I sat in the audience I almost felt sorry for the guy. Almost. It was painful to watch.

Then Caryl Phillips rose to say a few words. Phillips spoke softly and directly about the true city of Leeds that he knew from growing up in the part of Yorkshire where the black, brown and white working classes had lived, worked and played side-by-side. Leeds, he said, was and always had been a city of migrants and migrancy, of travellers and tradespeople. Leeds had been shaped over the centuries by these external flows of diverse peoples from across Europe and more recently from the Commonwealth, just as the River Aire flowed through the city giving it life and energy.

Leeds, Phillips told the audience, had always been multicultural despite what the inward looking monoculturalists would have us believe. Using a range of sporting analogies he showed how Leeds embodied a broader process of cultural exchange and translation across difference that had changed Britain herself and that this was what made Leeds strong. Its openness to strangers, its porous borders and boundaries alongside its sometimes fanatical pride in the local. His speech was beautiful, intelligent and moving. Delivered so subtly that its full force was undoubtedly lost on Illingworth. Each line, each sentence, each sentiment in what Phillips said was aimed directly at the worst of what Yorkshire had come to represent in the minds of many, while Phillips himself embodied and spoke to the best that Yorkshire could be. Illingworth just sat on stage with an expression that veered between confusion and discomfort.

Phillips sat back down to warm applause. The event came to a close. The VC seemed relieved. The cameras crews began to pack up. And Illingworth stood chatting to one of the other new degree holders Keith Hellawell, Chief Constable of West Yorkshire police. Phillips made some small talk and left the stage as soon as he politely could. Degree in hand, job done.

Closing Scenes: *Mise-en-Scènes*

'Scenes' have multiple meanings and iterations. The change from the signifier to the signified, from the visual rendering to its phonetic utterance further complicates its meaning. It suggests and invokes a place. A scene. A physical marker. Where black bodies are chased into canals. A death that was unseen. Or Asian bodies kicked down the same streets by trained, professional kickers of another sort. Crime scenes. Even if denied as such. Even if the attackers get off lightly. The streets know the truth. The streets: where black bodies are publicly searched. Looking for drugs where criminality is encoded onto the body. Where black bodies have to account for themselves: *Where have you been tonight?*

The physical place of the scene can also be a stage, a backdrop for performances of the physical and verbal kind. Performances of race both intended and unintended. A place too, where we can sometimes make a scene (of ourselves) in public. Scene as spectacle and scene as unintended disclosure, making visible (to be seen by all) that which is intended to be covered up, hidden behind the scenes. The racism that is kept off stage, in the back, but that sometimes wanders out by itself, unannounced. The props of racism. But the very act of attempting to conceal inadvertently reveals all. The racist speaking against himself: *I have lots of black friends*.

It can invoke the visual. To be seen. A form of surveillance and a controlling gaze. Chapeltown needing to be surveilled. The natives being watched with CCTV modern techniques. The Fanonion moment. The terrified look. Or more likely today, the curious look of bodies out of place, as though actors walking onto the wrong set, into the wrong scene. A young white girl watching a scene in black and white that doesn't make sense: *But you're black*.

Being stopped in the streets, by officers or by words themselves, hailing us into submission. The black body, as Fanon recalls 'surrounded by an atmosphere of certain uncertainty' (1986: 110-111). The physical can be dealt with. It hurts but often it can be overcome. The daily psychic abuses, big and small, constant, day after day, however cut deeper. To and into the bone. Sticks and stones may break our bones but it's the words that will probably kill us. Dying slowly in mental institutions over-subscribed with those the medical profession deem too mad, bad and black. The naming belies the impending violence: *negro, wog, nigger, paki*.

But to be seen can also be a space of active community and identity formation. Can the subaltern look back? A place to be seen thus becomes a scene. Here we are. We have arrived. We are not moving. We overlook the city and the city looks at us. Has to acknowledge our presence. A moment of recognition. It marks the limits of white terror. Draws a line. A boundary. Beyond which even white racism must fear to tread. This is our space now. Be warned. We no run. In fact, we score runs with panache: *Hold it, hold it!*

And finally, scenes can also be chapters, narrative devices, parts of a story. To be told. The opening and closing scenes of a film, or book, or an essay about a subject: Leeds. The story of Leeds continues to be told over and over again. This time with new characters. Asians as the new black. The new public enemy. Black down to number two (for now). Brown up to number one. Leeds imagined as the 'home' of the 7/7 bombers. Or at least three of the four. The Leeds lads who decided to exact a brutally violent and tragically misplaced political revenge on the innocent of London on that fateful morning of July 7th 2005. Suicide bombers play cricket too you know. Eat fish and chips. And study sport science at Leeds Met. How to make sense of that?

Stories cover over other stories that wait to be told. Need to be told. The sub-plots from the 1990s become today's lead story. How did we miss that? He's from London anyway so what would he know of Leeds. Who only Leeds know. This narrator struggles but fails to be a reliable guide. Scenes obscure as much as they reveal. Prevent us from seeing what is really going on. There is always a writer, a director and then an editor, first framing and then cutting the shot. Looking one way means not looking the other. Even the astute ethnographer, doctorate in hand, misses things. Important things. The unknown unknowns we might now say.

But all this does not mean we must not look. Just that we might try to look harder. Or at least to look in a different way, awry perhaps. To look more carefully at the ways in which the ideologies of race do their work, invisibly. Like Chapeltown itself on most maps. To trace the power geometry of race through the spaces where we walk, live and work and sometimes, some of us, if we miss the last bus home, or disrespect the wrong footballer, or meet the wrong officer on the wrong evening, die too. A critical reading of the city demands that we look with and from a different angle, especially at those accounts and histories that in romanticising the nostalgia of white northern-ness end up disavowing the racially complex present in order to correct the official colourless, distorted view. So that instead we can see more clearly. The various scenes and topographies of race. In Leeds as elsewhere.

Acknowledgements

For various responses, encouragements and inspirations, deep thanks to Caz, Chima, Grant, Lisa, Max, Neville, and Simone. Thanks too to the editors Pete and Steve for your editorial guidance, generosity and patience.

References

Ahmed, S. (2006). *Queer Phenomenology: Orientations, Objects, Others*. Durham: Duke University Press.

Aspden, K. (2008). *The Hounding of David Oluwale*. London: Vintage.

Back, L. (1994). The "White Negro" Revisited: Race and Masculinities in South London, in *Dislocating Masculinity: Comparative Ethnographies*, edited by *A*. Cornwall and N. Lindisfarne. London: Routledge.

Burton, R. (1991). Cricket, Carnival and Street Culture in the Caribbean, in *Sport, Racism and Ethnicity*, edited by G. Jarvie. London: Falmer Press.

Burton, R. (1997). *Afro-Creole: Power, Opposition and Play in the Caribbean*. London: Cornell University Press.

Carrington, B. (1998). Sport, Masculinity and Black Cultural Resistance. in *Journal of Sport and Social Issues*, 22 (3), 275-98.

Carrington, B. (1999). Cricket, Culture and Identity: An Exploration of the Role of Sport in the Construction of Black Masculinities, in *Practising Identities: Power and Resistance,* edited by S. Roseneil and J. Seymour. Basingstoke: Macmillan/Palgrave.

Carrington, B. and McDonald, I. (2001). Whose Game is it Anyway? Racism in local league cricket, in *'Race', Sport and British Society,* edited by B. Carrington and I. McDonald, London: Routledge.

Cohen, A. (1985). *The Symbolic Construction of Community.* London: Routledge.

Conn, D. (2006). Headingley Gropes its Way Toward Colour Blindness. *The Guardian*, March 22, 8.

Conn, D. (2008). The Beautiful North, in *Imagined Nation: England after Britain*, edited by M. Perryman. London: Lawrence & Wishart.

Fanon, F. (1986). *Black Skins, White Masks* London: Pluto Press.

Farrar, M. (1996). Black Communities and Processes of Exclusion, in *Corporate City? Partnership, Participation and Partition in Urban Development in Leeds*, edited by G. Haughton and C. Williams. Aldershot: Avebury.

Farrar, M. (1997). Migrant Spaces and Settlers' Time: Forming and De-forming an Inner City in *Imagining Cities. Scripts, Signs, Memory*, edited by S. Westwood and J. Williams, London: Routledge, 104-26.

Farrar, M. (2002). *The Struggle for 'Community' in a British Multi-Ethnic Inner-City Area: Paradise in the Making.* Lampeter: The Edwin Mellen Press.

Hall, S. (1996). A Conversation with C.L.R. James, in *Rethinking C.L.R. James,* edited by G. Farred. Cambridge: Blackwell.

Halpin, T. (2006). Lecturer is Suspended for "Racist" IQ Claims. *The Times*, [Online March 24] Available at http://www.timesonline.co.uk/tol/life_and_style/education/student/news/article694940.ece [Accessed : 2 November 2008].

Hoggart, R. (1957). *The Uses of Literacy: Aspects of Working-class Life with Special Reference to Publications and Entertainments.* London: Pelican.

Hunt-Grubbe, C. (2007). The Elementary DNA of Dr Watson. *The Times*, [On Line October 14] Available at http://entertainment.timesonline.co.uk/tol/arts_and_entertainment/books/article2630748.ece [Accessed: 2 November 2008].

Illingworth, R. (1999). Imran's Jibes are Way off the Mark, *Yorkshire Sport*, 21 June,13.

James, C.L.R. (1963). *Beyond a Boundary.* London: Serpent's Tail.

Lawrence, S. (1996). 'I'd have Chinned Illy if I'd Heard the Nig-Nog Slur', *Daily Mirror*, 13 November, 33.

Long, J., Tongue, N., Spracklen, K. and Carrington, B. (1995). *What's the Difference? A Study of the Nature and Extent of Racism in Rugby League.* Leeds: The Rugby Football League/Leeds City Council/The Commission for Racial Equality/Leeds Metropolitan University.

Long, J., Nesti, M., Carrington, B. and Gilson, N. (1997). *Crossing the Boundary: A Study of the Nature and Extent of Racism in Local League Cricket.* Leeds: Leeds Metropolitan University Working Papers.

Lynskey, D. (2007). Readers Recommend: Songs about Northern England, *The Guardian, Features Section*, 7 September, 4.

Maguire, C. (1997). 'CRIME FEAR SHUTS BANK: Community anger at blow to suburb', *The Yorkshire Evening Post*, 16 January, 1.

Marqusee, M. (1994). *Anyone but England: Cricket and the National Malaise.* London: Verso.

Massey, D. (1994). *Space, Place and Gender* Cambridge: Polity Press.

O'Hagan, A. (2006). The Future Sounds Rosier with Hovis. *The Daily Telegraph*, [Online, 3 May) Available at http://www.telegraph.co.uk/opinion/main. jhtml?xml=/opinion/2006/05/03/do0303.xml [Accessed: 2 November 2008].

Phillips, C. (2002). *A New World Order: Selected Essays.* London: Vintage.

Phillips, C. (2005). Northern Soul. *The Guardian: Weekend*, 22 October, 18.

Phillips, C. (2007). *Foreigners.* New York: Alfred A. Knopf.

Policy Research Institute (1996). *Community Profile of Chapeltown Leeds.* Leeds: Leeds Metropolitan University.

Searle, C. (1990). Race before Wicket: Cricket, Empire and the White Rose, *Race & Class*, 31 (1), 31-48.

Searle, C. (2001). Pitch of Life: Re-reading C.L.R. James' Beyond a Boundary, in *'Race', Sport and British Society*, edited by B. Carrington and I. McDonald, London: Routledge.

Taylor, M. (2006). University Suspends Lecturer in Racism Row who Praised BNP, *The Guardian*, 24 March, 5.

Thompson, E.P. (1963). *The Making of the English Working Class.* New York: Vintage.

Chapter 7

Nowt for Being Second: Leeds, Leeds United and the Ghost of Don Revie

Stephen Wagg

Strange that the mind will forget so much of what only this moment is past and yet hold clear and bright the memory of what happened years ago – of men and women long since dead. For there is no fence nor hedge round time that is gone. You can go back and have what you like of it.

Philip Dunne *How Green Was My Valley* (1941)

It's all right not to believe in luck and omens. Nobody believes in them. But it doesn't do any good to take chances with them and no one takes chances. Cannery Row, like every place else, is not superstitious but will not walk under a ladder or open an umbrella in the house.

John Steinbeck *Cannery Row* (1945)

During early evening on 22nd May 2008, as Premiership clubs Manchester United and Chelsea prepared to contest the final of the Champions League in Moscow, BBC Radio 2 disc jockey Chris Evans began in the course of his *Drivetime* programme to muse on the fact that these two clubs carried a great deal of debt. In a subsequent telephone conversation with Evans, Professor Stefan Szymanski, a specialist in football finance from Cass Business School in London, explained that the two clubs had borrowed heavily in anticipation of economic growth. Football clubs should borrow more, enthused Prof. Szymanski, and this would make the game more exciting, because some of them would 'come crashing down'. 'I mean', he added in support of his argument, 'look at Leeds United'. The fallen Leeds United meanwhile would three days later meet historically greatly less celebrated Yorkshire neighbours Doncaster Rovers in the play-off game for promotion to the Championship – effectively, England's second division. They lost.

This chapter is about this apparent fall from grace, if only in the specific sense that it examines the enduring importance in the life of the city of Leeds teams of the 1960s and early 70s. Of course the personnel of the Leeds United team changed greatly over a decade, but it is remembered, in effect, as one team, along with a small subsidiary group of precedents and replacements. Most remembered of all is probably the team's manager throughout this period, Don Revie. I want here to examine the meanings of Revie's teams in the football world, both in Revie's time and since, and in contemporary Leeds, where they are the basis of a mini-heritage industry. I also wish to discuss the Leeds United of the last two decades or so (1990 to the present) and I will try to make sense of this more-recent Leeds United in relation both to Revie's Leeds and to the development of Leeds as a postmodern city.

Leeds United and Don Revie: A Sociology of 'Dirty Leeds'

On 4th June 2008, BBC Sport reported that the England shirt worn by Don Revie in his first international match – against Northern Ireland in Belfast in October of 1954 – had been stolen from the reception area in the West Stand at Leeds United's Elland Road stadium. United were appealing for help in finding this 'valuable piece of the club's history' (BBC Sport, 2008). Strictly speaking, this shirt was not part of the club's history at all, since Revie had been a Manchester City player when capped for England, but the mistake was understandable. So powerful is the myth of Don Revie as Leeds United's creator that the period from the club's foundation in 1919 to Revie's appointment as team manager in 1961 is described by the biographers of the Revie team as 'Leeds pre-history' (Bagchi and Rogerson 2002: 25).

Paradoxically, Revie is, in part at least, remembered for not being remembered. Bagchi and Rogerson note that when the ex-Manchester United manager Matt Busby died in 1994 a minute's silence was observed on all English league grounds, with the exception of Blackburn Rovers' Ewood Park, where Leeds United were due to play and hundreds of their supporters chanted 'There's only one Don Revie' throughout (Bagchi and Rogerson 2002: 4-5). Similarly, although Revie had managed the England team between 1974 and 1977, when he died in 1989 no representatives of the football authorities attended his funeral (Bagchi and Rogerson 2002: 205). In life and in death, Don Revie and his Leeds United have aroused strong emotions in the football world – emotions recently revealed by the publication in 2006 of *The Damned United* (Peace, 2006), David Peace's novelised account of the 44 uncomfortable days spent at Elland Road by Revie's successor Brian Clough. In an edition of ITV's *The South Bank Show* devoted to Peace's book (11th May 2008), the journalist and former professional footballer Eamon Dunphy described Revie's Leeds as 'a boil of the face of the game', adding that Clough, whose open endorsement of this broad view had made him an unlikely replacement for Revie, had said openly what 'many people in football believed but daren't say'. Clough would later write of the Revie team that they were 'one of the most cynical and dirtiest as well as talented I had ever seen' (Clough and Sadler 2005: 174).

Much of the discourse about Revie, Clough and *The Damned United* during 2007-8 has been about the personalities of the individuals involved: one Leeds player – John Giles – successfully sued for libel over his portrayal in the book and *The South Bank Show* carried a statement from Clough's widow saying that in life he had not resembled the man depicted in Peace's novel. Here, though, I want to try to explain Revie, Clough and the Leeds United sides of the 1960s as people of their time, dealing with and reacting to important social changes, within the football world and beyond. Revie and Clough had importantly complex relationships not only to the Northern working-class culture that had spawned them both but also, within football culture, to modernity (the rationalisation of the game and its progress toward *technocracy* – the regime of experts) and to postmodernity – the growing acceptance of football as a commodity, a brand, a television show.

Revie, like Clough, came from a working-class home in Middlesbrough, an industrial port in the North East of England and, at the time, part of the North Riding of Yorkshire. He was born in 1927 and his father, a joiner, was unemployed for two years while he was young. His mother often took in washing from a more affluent area of town (Mourant 2003: 13) – a classic signifier of poverty in inter-war Britain. Revie himself was apprenticed as a bricklayer. As a player he was accomplished, holding six England caps and latterly, at Manchester City, becoming known for pioneering in English football the strategy of 'deep lying centre forward' – a tactic apparently inspired by the successful Hungary team of the early 1950s. This helped found the reputation that Revie was to enjoy as a technocrat open to continental influence.

Revie became player manager of Leeds United in 1961. In what is, arguably, the most eloquent defence of Revie's Leeds, Rob Bagchi and Paul Rogerson paint an evocative picture of the city in which Revie sought to establish a successful football team. Leeds was the headquarters of Yorkshire County Cricket Club – the county's sporting flagship – and home also to three Rugby League clubs: Bramley, Leeds and Hunslet. During Revie's early years at Elland Road, the board of directors talked intermittently about wanting 'to keep soccer in Leeds' (Bagchi and Rogerson 2002: 40) and complained when attendances were low – 'The Leeds public disgust me' said chairman Harry Reynolds when only 17,753 people showed up to watch Leeds play Rotherham on the first day of the 1961-2 season (Risoli 2003: 172). The industries that had supported the city's growth in the nineteenth century were, they point out, already in decline in 1961 and the service industries and the '24 hour city' culture for which Leeds is now noted, were already in evidence. The consumer society was also well established with three quarters of Leeds households owning television sets. Unemployment in the city was less that 1% (Bagchi and Rogerson 2002: 6-13). To a degree, the club's directorate reflected these changes: Harry Reynolds, Revie's first chairman, was a millionaire former railwayman who'd made his fortune in scrap metal, while two of his immediate successors Albert Morris (1967-8) and Manny Cussins (1972-83) had prospered through the consumer boom of the 1950s and 60s, Morris in the wallpaper business and Cussins was chair of the furniture firm Waring and Gillow. Like most directors of the time, Reynolds, Morris and Cussins were prominent figures in local commerce, perceiving themselves to be stewards of the people's game: Reynolds' daughter, for instance, told Revie's biographer that 'when he retired in 1959 Leeds United became more or less his hobby' (Mourant 2003: 55).

They would, from time to time, support Revie's team building, occasionally with interest-free loans out of their own pockets (Bagchi and Rogerson 2002: 40).

Revie's managerial regime is widely written on and, in its fundamentals, broadly agreed. On the field of play Revie applied, effectively, Fordist principles: each Leeds team (first, reserve and youth) were tutored in the same system of play and each individual allocated a specific role within it, carefully explicated by Revie himself and administered by his coaching staff (Bagchi and Rogerson 2002: 20). This system is probably best explained by Jim Storrie, one of Revie's early signings:

> We were a very physical, hard-working and hard-running side. It was high-pressure football. We had to put the opposition's players under pressure all over the park. We harassed and chased. Revie was one of the first managers to introduce that way of playing. I was a forward and my first job was to defend. That was the mentality. (Risoli 2003: 179)

Two elements in this strategy were controversial in the football world of the 1960s and 70s. One was the degree of *prescription* entailed in Revie's approach (he often issued his players with dossiers, detailing supposed strengths and weaknesses of their opponents) and the other was the perceived tendency toward intimidation and devious practice that inspired the soubriquet 'Dirty Leeds'.

On the first count, Revie was simply embracing a broadly technocratic stream of thought that had been around in professional football circles for much of the twentieth century and embraced by the FA since the 1930s (Wagg 1984). Coaching had been accepted on the continent of Europe for much of that time and club managers – notably Rotherham-born Herbert Chapman, who had managed Leeds City (1912-19) and Huddersfield Town (1921-1925) before moving to Arsenal – was an early proponent. There is much evidence to support the claim that Chapman was English football's first and perhaps most influential modernist (Say, 1996, Studd, 1981) – not the least of which is his own *Chapman on Football*, published in 1934, the year of his death (Chapman, [1934] 2007). Chapman, it's clear, was implementing modernist ideas well before the First World War at Northampton Town, where he went as player manager in 1907.[1]

Most importantly, these notions of modernism and technocracy in football were not reducible to social class differences; rather, despite prevailing ideological conflicts in the football world, they cut across these and other, crucial social differences. Thus, historically, there were men in amateur football circles who disdained any form of tactical preparation and, equally, men, notably those who taught or had trained at one or other of the leading physical education colleges established in England in the 1930s, who happily dealt in football theory.

---

1   Although widely discredited the Wikipedia website often provides excellent accounts of people and/or issues. Its entry on Chapman is, I think, one of them: http://en.wikipedia.org/wiki/Herbert_Chapman. My access: 30th June 2008.

Likewise, for every Herbert Chapman, there were, for much of the first half of the twentieth century, innumerable professional footballers who thought that football was something that couldn't be taught or usefully rehearsed: it was the game of the people, growing organically out of the virtuosity of the working class. This latter view had different manifestations at different historical moments. It could, for example, be readily mapped on to the historic North-South divide, the assumption being that the South of England represented 'toffs', amateur footballers and the resented Football Association, while the fabled North (Russell, 2004) was the place where professional football was born and where working-class boys had learned the game, as Don Revie himself had done (Mourant 2003: 14), kicking a ball of rags around a cobbled street. The two competing philosophies were equally marriageable with prevailing ideas of masculinity: for working-class males, to receive coaching could be seen alternatively as effete or as part of the work of a 'tough professional'. All this can help explain the apparent hostility, and the ultimate complexity, of the response to Revie's managerial regime.

At the time Revie was finding his way as a football manager, important changes were taking place. In 1961, the year he got the Leeds job, the maximum wage restriction for professional footballers in England was abolished which meant that clubs would now have to look more closely at their overheads and their records of achievement. Football was also becoming a television show with the inauguration of the BBCs Saturday evening highlights programme *Match of the Day* in 1964. Moreover, a muted ideological war was being fought over the issue of technocracy itself. This purported tussle between 'Muddy Boots' and 'Chalky Fingers' seemed to spring out of the threat League football people perceived to their jobs from FA-certificated coaches who'd 'never played' (i.e. professionally). They may have the paper qualifications, ran the standard argument, but, in the real world of winning football matches, what did *they* know? (Wagg 1984: 73-100).

Revie appears to have got the idea for his (in)famous dossiers, and the ethos that went with them, not from an FA course, but from his experiences in working-class football in Middlesbrough. They are said to have been the brainchild of Jim Sanderson, a train driver and part-time coach of a youth team, who held team meetings in his council house at which detailed breakdowns of opposing teams were handed round (Corbett, 2007). When Revie introduced these and related ideas to the Leeds United dressing room, they were widely labelled 'professionalism'. Some occupants of this dressing room were uncomfortable with them. 'You'd be sitting there thinking: 'God, just let us play'', recalled Mike O'Grady, who played for Revie between 1965 and 1969 (Mourant, 2003: 101). Others welcomed these methods. John Giles, who signed from Manchester United in 1963, said: 'I found the first day I went to Leeds an atmosphere I hadn't known at Manchester United. There was a buzz about the place, a keenness, a will and an attention to detail …' (Bagchi and Rogerson 2002: 68). Norman Hunter (Leeds United 1962-1976) offers a widely favoured defence of Revie's instructions – that the manager was simply a modernist ahead of his time. Referring to the tactic of taking the ball to the corner flag and challenging the opposition to come and get it, Hunter recently argued: '...we were lambasted by the media for it. Time wasting is accepted as part of the game now' (Hunter and Warters 2004: 84). This disapproval, though, would not have been confined to the media: in the English football world all this would represent is a cultural shift. Jimmy Greaves, doyen of untutored post-war English footballers, has stressed that his club, Chelsea, League Champions in 1955, still had no coach in 1959. 'The joke within the club was that Chelsea were the 'all the best' team, because invariably that was all Ted [Drake, the manager] would say to the team before the game .... Ted's hands-off approach to management was in fact typical of many managers of this time [...] When I started out [in 1957] no one gave a monkey's how many goals you conceded, the aim was to score more than the opposition' (Greaves 2003: 41, 45). The Revie way, introduced only four years later, ran strongly counter to this of course. Billy Bremner, Leeds captain through much of the Revie era, reflected: 'When we were away from home and we scored a goal, I can remember thinking that the 25,000 people or so watching would be as well going home there and then ....' (Mourant 2003: 84). Bremner, though, argued that such methods, while enjoying only minority support in England, were readily embraced elsewhere in Europe. Revie's teams were influenced by Italian sides and admired in that country (Bagchi and Rogerson 2003: 99-100). As Bremner remarked, 'We thought a lot about our game and picked up traits from the Continentals. What we called cynical in this country was called professionalism when the Italians played it' (Mourant 2003: 85). Too close a comparison between Leeds and Serie A was called into question, however, when Revie signed the Welsh international John Charles from Juventus of Turin in 1962. In a piquant early marriage of the local and the global, Charles, the epitome of European football sophistication, was welcomed with a fish-and-chip supper at The Railway Inn in Beeston, a working-class district encompassing Elland Road itself (Risoli 2003: 170). But, according to Jim Storrie, John 'wanted to play one-touch football and flick the ball here and there. At the time that wasn't Leeds' style. Long balls were played to the corner flag and John was expected to chase after them. At half-time in one game I remember John saying, "I'm not running my pants off for long balls". And wee Billy Bremner said, "You're making that fuckin' obvious"' (Risoli 2003: 180). The Charles experiment notwithstanding, the reputation of Revie's Leeds grew in Italy through the 1960s and, in May of 1969, he was offered the managership of the Italian club Torino (Mourant 2003: 109-10).

A related issue was, and remains, the matter of the physical nature of Leeds' play under Revie as entailed in their enduring description in many quarters as 'dirty Leeds'. 'It is, I think, generally understood', wrote *Times* columnist Giles Smith in 2006, 'that, in any argument about right and wrong in football, a reference to Revie's Leeds United is the nuclear option. There is, quite simply, nowhere to go after that. There has never been a more horrible football team ....' (Smith 2006). There is, I think, no point in trying to determine what justification there might be for this remark. It's important, though, to understand the complex way in which the idea of physical aggression was sanctioned in the Revie project.

First, there's little doubt that a number of Revie's players were, or have become, disquieted about the way they were required to play. Bagchi and Rogerson, in their impressive 'redress-the-balance' book on Revie's Leeds, quote Eddie Gray (Leeds United 1965-83) as saying (in his autobiography) that he would sometimes 'wince at our approach' and John Giles as 'now thoroughly ashamed' of the 'bad things' he did in a Leeds shirt (Bagchi and Rogerson 2002: 72, 140). In his memoir of 1996, dedicated to Revie, Jack Charlton (Leeds United 1952-73) recalls Jimmy Lumsden, a young Leeds reserve player, boasting 'that he had gone in over the top of the ball to a guy who then had to be taken off'. 'Don', adds Charlton, 'murmured something approvingly' (Charlton and Byrne, 1996: 66). More recently, Terry Yorath has described playing for Leeds in a youth team game in 1966. Leeds needed to beat Hull in the final game of the season to win their league and Yorath was instructed by club captain Bobby Collins to 'take the Hull winger out':

> My concern was just to get him off the park. The ball got thrown to this lad and I just went whack! Straight into him. He was carried off on a stretcher with a broken leg – although I didn't know it at the time – and we ended up winning the game. [....] I still see Bobby Collins quite a lot these days and he always reminds me of the incident: "Remember that tackle", he says. "That won us the league". I admit it's not a nice story but it just shows how the competitive spirit under Don Revie had affected everyone at Leeds. (Yorath and Lloyd 2004: 28)

The considerable writing on Don Revie's Leeds shows that he and his managerial staff developed an ideology around the Leeds team which, among other things, rationalised any misgiving about the way the players played. At the heart of this ideology was Revie himself as *paterfamilias*. This notion of manager-as-father-figure was, in itself, not unusual for the time. With the removal of the players' maximum wage restriction in 1961, some managers began to move closer to their teams, fussing over their personal needs, appearing in tracksuits at training, and so on (Wagg 1984). 'I would walk through walls for my players' Revie told me when I interviewed him briefly as a postgraduate student in 1972. Part of the logic of this was that conspicuous attention to players' non-material welfare might mitigate their financial demands: footballers might feel uncomfortable about asking for a higher wage from a manager who seemed more like a father or favourite uncle than an employer. 'Going in to see Don Revie about a pay rise was not something any of us looked forward to ....' recalled Norman Hunter. 'All of a sudden, it was you and him. You would come out having got what he wanted you to have. You found yourself saying "Thanks very much, Gaffer"...' (Hunter and Warters 2004: 171). And when Allan Clarke signed for Leeds in 1969 'Don asked me what wages I wanted. I told him I would like £10 more than I was on at Leicester. He refused and told me all his players were on the same wage. Who was I to disbelieve him?' (Saffer 2001: 62).

In return for this reluctant wage restraint, Revie offered what was, in effect, a pathway through a new social and economic universe. Rather in the manner of the Canadian social psychologist Erving Goffman's 'total institutions' (Goffman 1991) Revie's staff sought to fashion their players' identities, isolating them where possible from certain influences and accentuating others. The 1960s saw the beginnings of a free market in English football. This would bring – indeed, was already bringing – greater rewards for players, heightened competition and growing media scrutiny. Revie's Leeds offered a strategy for these new times, drawing heavily on established conventions of social class, region and gender.

Revie, as we've seen, was a Northern working-class male, born in the 1920s. In the society of his time, class and status ordinarily went together, the chief implication of which was that if a person made money and became upwardly mobile, they had to change their culture accordingly. British situation comedy, for example, is littered with characters who tried, unsuccessfully, to 'talk posh'; so, for that matter, is the history of English sport – the Leeds-born England cricket captain, Sir Len Hutton, for example, and England football manager Sir Alf Ramsey come to mind. Revie grappled with these same difficulties, agreeing with some reluctance to send his children to boarding school (Mourant, 2003: 112). In his acceptance speech as Manager of the Year in 1970, he applauded his managerial hero, the disciplinarian Willie Struth (manager of Glasgow Rangers 1920-54), for sending any player whose hair touched his collar 'to attend upon the hairdresser' – an archaism, as Bagchi and Rogerson note, even for the time (p.139). In this regard, Revie sought to shepherd his players toward a set of middle-class proprieties that were, arguably, ceasing to exist. During the 1960s he and the players moved from inner-city districts like Moortown, Beeston and Wortley either to the affluent suburb of Alwoodley in North Leeds or to equally affluent towns nudging the dales north of the city. Revie took them away from their families before home matches, putting them up in a hotel in Ilkley and often sending their wives flowers in recompense. Here and on other occasions he encouraged them to inhabit a world redolent of the Yorkshire 'dalesman' lower middle class of the 1930s, wherein they played parlour games and carpet bowls – a world considerably more J.B. Priestley than Mick Jagger. In the (elsewhere) 'Swinging' 1960s Revie had his players singing songs like 'Ilkley Moor Ba' Tat' – an 'Elland Road' version of which they sang after they won the Inter Cities Fairs Cup in Budapest in 1968.[2] Of the leading players of Revie's Leeds only Paul Madeley (born in Beeston in 1944) was actually from Leeds; during the 1960s, though, they became apprentice 'dalesmen'. This acquired Northern identity, for which Yorkshire County Cricket Club was traditionally the standard bearer, became a crucial part of their armoury when dealing with criticism of the 'Dirty Leeds' variety. Billy Bremner, although a Scotsman, born in Stirling in 1942, nevertheless titled his autobiography with the terse Yorkshire-ism *You Get Nowt For Being Second* (Bremner 1969). From an entrenched position in this mythic 'North' where life-was-tough, criticism was more easily disposed of: it became assimilated to the equally mythic North-South

---

2   A story confirmed for me by Rob Bagchi in a telephone conversation 10th June 2008. Rob was told it by Jack Charlton.

divide wherein honest northern toil faced down southern pretension – in the form of the F.A., whose published disciplinary records Revie perceived to be part of a campaign against Leeds United (Bagchi and Rogerson 2002: 81) or the London-based football press with its preference for 'fancy' teams such as Chelsea and Tottenham Hotspur. The manager, remembered Norman Hunter, liked to use press criticism to the team's advantage: 'Don Revie would bring in the cuttings to show us ….. "Look what they're writing about us now", he'd say. "We'll show 'em"' (Hunter and Warters, 2004: 55). 'Playing London teams was always competitive', recalled Paul Madeley (Leeds United 1963-80) 'partly because the national press gave us a hard time. It did aggrieve us because we had some fantastic players' (Saffer 2003: 72).

Within the Leeds 'family', equally strong notions of gender ran alongside class and region. In the decade of the contraceptive pill, sexual exploration and nascent feminism, Revie wished his players to be married, with a home to go to. When Terry Yorath signed for Leeds as a full professional in March 1967 'the manager told me I'd done well and then gave me a lecture about the right way to live. He stressed the importance of having good digs and, as he did with all players, he later encouraged me to get engaged as soon as possible' (Yorath and Lloyd 2004: 29). Conversely, Mike O'Grady, who left United in 1969, was convinced that Revie let him go because he was the only first team regular that was single (Mourant 2003:111). With their female helpmates at home, awaiting either their key in the door or a compensatory bouquet from the boss, the Leeds players were encouraged to be tough, disciplined professionals at work. Revie's coach, Syd Owen and his trainer, Les Cocker combined technocracy with a stern, barrack-room masculinity. In this environment aggression and self defence merged into one and were sanctioned by various successful appeals – to group solidarity, to professionalism and to the need to be safe in a dangerous world, one in which it might be necessary to 'get your retaliation in first'. This culture is unlikely to have been unique to the Leeds United of the 1960s, but it does seem to have been writ large there. Peter Lorimer said in 2007: 'Our whole ethos was built on loyalty. We all fight for each other, we all work for each other. If someone kicks me, he kicks all 11 of us' (Corbett, 2007). Norman Hunter remembers: 'Les Cocker would play in the practice games and he used to clatter into you, sending you sprawling. Often, you would end up hitting the fencing. I think it was his way of trying to toughen us up. On one occasion, shortly after I made it into the first team ….. I charged in and hit him with everything I had. I put him into the air and into the fencing. I wondered what I had let myself in for but Les looked up at me with a huge smile, as if to say, "Great, you've learnt, you've got the idea now …."' (Hunter and Warters 2004: 38-9). Pete Rodgers, who played for Leeds United youth team between 1964 and 1966, said recently that one day Cocker put up a chart showing the vulnerable, most easily injured parts of the body: 'He never told us to kick these parts. He said this was in order to defend ourselves. But I think some of the lads thought "This could give us the edge"'.[3] In this respect, then, Cocker and Revie seem to have caught the mood of the times, as well as helping to define it. 'The 1960s' reflected John Giles in 2005, 'were a hard time ...' (Doyle 2005).

In the summer of 1974, into this harsh, disciplined world of bonded males, strolled Brian Clough, the most outspoken critic of Revie's Leeds and now, improbably, their new manager. Joe Jordan (Leeds United 1970-8) observed: 'Clough came into Elland Road wearing shorts and a T shirt and holding his son Nigel by the hand' (Jordan and Lawton 2004: 91). Syd Owen told Revie's biographer of his disgust at the manner of Clough's arrival:

> The first morning Clough came to the club as manager, he brought his two sons with him. I was just taking the reserves and the junior players out for training when he shouted to me down the corridor and asked if I could get one of the apprentices to take his two boys into the gymnasium and entertain them while we were out. I told him the apprentices were here to develop their capabilities as professional footballers and not look after the manager's boys. I said: "There are no groundstaff boys available. They are not here to entertain your children". After that I would just get on with my job and avoid him. (Mourant 2003: 155)

Of course, the reasons for this unfriendly welcome given to Clough at Elland Road went well beyond his apparent profaning of a proud, emotionally-contained adult male work environment with casual clothes and small children. There is known to have been a strong antipathy between Revie and Clough and Clough's expressed philosophies seemed to run counter to 'Revie-ism' on virtually every count. However, although in obvious conflict at the time, the two men in their respective ways represented the face of English football's future.

Clough, like Revie, had been born in working-class Middlesbrough and was by eight years' the younger man. His father worked in a sweet factory. Although he played briefly for England (two caps in 1959) under their first coach Walter Winterbottom, a leading apostle of technocracy, Clough himself was never associated with the coaching movement. Indeed, he disdained coaching and extravagant tactical planning. A biographer would later write: 'Team talks [for Clough and his assistant Peter Taylor] were brief and uncomplicated. There were no thick dossiers on the opposition, no blackboards (or 'blackbores', as Clough called them), no diagrams to follow, no fretting about what tactics the other side might use' (Hamilton 2008: 51). Indeed, Clough's inspiration in football management was said to come from Alan Brown (manager of Sunderland 1957-64, 1968-72) and Harry Storer (manager of Derby County 1955-1963), both managers whom he had encountered as a player in the 1950s and known disciplinarians. Storer in particular was known to be a sergeant-major figure, wont to bark sarcastically

---

3   Contribution from the floor at Anthony Clavane's talk 'Hubris, Schmubris: The Plot Against Leeds United' Institute for Northern Studies, Leeds Metropolitan University, 11th June 2008.

at players for their lack of competence or courage (Hamilton, 2008: 55) – an approach thought increasingly inappropriate in the 1960s. Yet, in 1972, Clough, with methods (or a lack of them) which seemed to belong to a bygone age, had won the League Championship with Derby County, a small-town club with modest resources, beating Leeds into second place.

This made Clough the perfect embodiment of the now-emergent myth of the modern football manager as the sole architect of a team's victory. Here was a man who seemed to marry the old world and the new. He cheerfully accepted responsibility for the team's performances, but did not dull those performances with science or defensiveness – apparently, he just told the players to go out and play, like in the 1950s. As a popular pundit on TV and in the press, Clough castigated Revie's Leeds, among other things, for their lack of spontaneity. In his autobiography, while acknowledging their talent, he writes: 'I had the impression that Leeds could have been more dazzling still had Revie been less systematic and allowed them off that tight rein of his' (Clough and Sadler, 2002: 175). He also mocked the Leeds family ethos as 'having more in keeping with the mafia than *Mothercare*' (Corbett 2007).

But, if he called up the 1950s in his managerial invocations, Clough was a man of his time in his relationship to the wider British society of the 1960s. In this society, football was becoming popular media culture and so, in a particular rendition, was the North of England. Novels such as John Braine's *Room at the Top* (1957), Keith Waterhouse's *Billy Liar* (1959), *Saturday Night and Sunday Morning* (1958) by Alan Sillitoe and *A Kind of Loving* by the Wakefield novelist Stan Barstow (1960) all portrayed restless working-class males railing against the financial and social constrictions of northern small-town life – constrictions not dissimilar to those that Don Revie was trying to impose on his players.

In particular, Arthur Seaton, the hero of *Saturday Night* ..., which is set in Nottingham, wants a good time on his own terms (Sillitoe, 1960; Samuel, 1998: 164). The ideal held out to such young men by popular culture and political conventions of the 1960s was that it was possible to progress financially and still have fun and, crucially, *be yourself*. Clough, as the novelist David Peace noted, 'seemed to have stepped out of these novels' and was known to have loved both the book and the film of *Saturday Night and Sunday Morning* (*South Bank Show* ITV 11th May 2008). Where Revie's teams pursued an other-directed philosophy, Clough's Derby had been content to let the opposition worry about them. Similarly, while Revie concerned himself with proprieties, coaxing his players into marriage, parlour games and membership of the local golf club, Clough seemed to swing the world by the tail, speaking as he pleased, apparently careless of the consequences. He wanted to speak on his own account, and not via the perceived conventions of a social class into which he might have migrated: 'Clough – he's the crazy one, isn't he?', said a European journalist years later to Clough's biographer (Hamilton 2008: 115).

Revie was famously superstitious, thus marrying a hard-headed modernism to a pre-modern belief in fate. He once had a 'gypsy' come over from Scarborough to lift what he felt was a curse on Elland Road (Mourant 2003: 91) and was known besides for his lucky suits, fear of birds and a range of anxious match day rituals. Clough cheerfully contended that, in matters of football management, he himself provided the magic – luck didn't enter into it: his autobiography, first published in 2002, the year before his death, is subtitled 'Walking on Water'.

While Revie enjoyed a good working relationship with Leeds directors and went to live among them, Clough often publicly derided his directors, declaring, like Storer, that they knew nothing about football (Hamilton 2008: 54).

Unlike Revie, Clough had no wish to marry class with culture. He did not speak with a recognisably Middlesbrough accent, nor did he embrace overtly middle-class conventions; he seemed instead to have reinvented himself, with a slightly whining, declamatory voice that belonged, geographically, nowhere in particular. He transcended the North and the working class and belonged instead to television, to the media. Revie, who in the oft-quoted words of Arthur Hopcraft, was 'a big, flat-fronted man with an outdoors face as if he lives permanently in a keen wind' (Hopcraft 1968:103), did not.

Both men, on the other hand, had some understanding of coming of the postmodern in football – the increasing centrality of television and the growing importance of 'image'. At the beginning of his managership, Revie had decreed that Leeds should wear all white kit. This was partly to try to capture some of the glamour that attached to Real Madrid, the premier club side in Europe at the time and participants in a thrilling European Cup Final in 1960 in which they beat Eintracht Frankfurt 7-3, widely seen on television. (And it was partly to provide a focus for his players' aspirations: Europe). But, ultimately, television and the media generally, were, for Revie, part of the outside world which was unlikely ever to give him or his team a fair shake. He is instead remembered as one of post-war English football's most influential technocrats and modernists. As Corbett argues, 'Revie changed the face of English football. He was a confidant to the players, psychologist, social secretary, kit designer, commercial manager, PR flak, dietitian and all-encompassing 'boss' of his team. In an era when pre-match preparation consisted of a 10-minute chat before a game, Revie was a revolutionary' (Corbett 2007). Clough, on the other hand, improbably disdained modernism but embraced the postmodern world of the television studio, the sound bite and the press conference: he was, arguably, English football's first celebrity football manager.

Goldfish and Money to Burn: Leeds United and Globalisation

Revie and Clough each, as managers, twice won the English League Championship: Revie in 1969 and 1974 and Clough in 1972 and 1978 – with different clubs, the only post-war manager to do so. The clubs involved – Leeds, Derby County and Nottingham Forest – all currently lie outside of the top division and few serious students of the game can think that any of them could repeat the feat. This, broadly speaking, is because of globalisation and I'd like briefly to consider Leeds United's more recent history, in relation both to globalisation and Don Revie's legacy.

Globalisation means, simply, the decreased importance of national boundaries. This can apply to variously important matters such as governance, commerce and culture. In the case of football, it has manifested itself in the increased significance of F.I.F.A., the governing body of world football and regional authorities such as U.E.F.A.. For Revie, as for most of his fellow managers at First Division clubs in the 1960s 'getting into Europe' was the ultimate objective. For a leading English club today, as Leeds United's experience shows, it is prerequisite.

After Revie and Clough, Leeds United for the most part struggled to retrieve past glories and to reassert former identities. Clough's immediate successors were high-profile men from out of town, notably Jock Stein, the Glasgow Celtic manager whom Revie had greatly admired (Bagchi and Rogerson 2002: 139). Stein, as it turned out, lasted no longer at Elland Road than Clough: 44 days. But the talismanic quality of the 1960s team in the West Riding area was reflected in the fact that, after 1980, four of Revie's top players were invited back to manage the club: Allan Clarke (1980-2), Eddie Gray (1982-5, 2003-4), Billy Bremner (1985-8) and, in addition, Norman Hunter, who was caretaker manager for two weeks on Bremner's departure. Much of the 1980s (1982-90) were spent in the Second Division, with the board apparently hoping that these figures from the glory days could rekindle the enthusiasm of players and supporters.

Bremner's successor, Howard Wilkinson, a Yorkshireman born in Sheffield, asked Hunter to leave, telling him 'that he didn't want any Leeds players on his staff' (Hunter and Warters, 2004: 180). 'I was aware', Wilkinson explained later,

> that that was all the club had – memories. I felt that too often those ghosts were comforters at the club and I thought that they should be locked up, put away until we had a new club and a team worthy of those reminiscences. So, one of the first things I did was take all the old [Revie era] pictures down .... (Moynihan, 2007: 201)

Wilkinson's Leeds gained promotion back into the First Division in 1990 and two years later won the First Division title, the last team to do so. It's fair to say, though, that the Wilkinson side is not remembered in the Leeds area with anything like the affection accorded to the Revie team. This may be partly to do with Wilkinson's well-meaning renunciation of Revie's legacy. It may also be that, Wilkinson, a quiet, undemonstrative man in the Revie mould, was not glamorous enough to be remembered for long in the television age. Speaking of which, the English Premier League, linked to huge sponsorship and satellite television coverage was inaugurated the following year, simultaneously designating 1992 as the end of an era.

In any event, in his book of 2002, David O'Leary, who became manager of Leeds United in 1998, makes no mention either of Wilkinson or his team when he reflects on how he approached the job. Here, once again, the talk is of Revie:

his legend lived on at the club as a man of vision and accomplishment, but 'if we are totally honest, we have to accept that Revie's was not a team that won the hearts and minds of the sporting nation'. Words, he continued, 'like "cynical" and "brutal" are never far from any discussion of the Revie years among unbiased football people'. 'I had a vision of a new Leeds team that would delight real football fans all over the country. I wanted to win over the neutrals. I wanted the television companies to feel they were assured of an entertaining programme when they covered our matches' (O' Leary 2002: 2).

Two things must be noted here. The first is the central importance of television – a vital factor in the 1990s but no more than a bonus, even an intrusion, in Revie's time. A successful team now would have to have broad appeal – appeal that stretched outside the city, outside the North, outside Britain. Second, it would cost a good deal of money – more than Don Revie or the benevolent local businessmen on his board of directors could have dreamt of – to assemble.

League football was now fully reconstituted as a business. By now, Leeds, like many other clubs, were no longer run by the sort of men who ran them in Revie's day – local businessmen pursuing an enthusiasm married to a sense of civic duty. At Elland Road the key changes came in 1996 when the club was sold to the London-based Caspian Group, in the process greatly enriching, by £2 million each, the three directors who were the major shareholders and who had previously formed a holding company to run the club: Leslie Silver, Peter Gilman, a local property developer and Bill Fotherby, who was in the textile trade (Rostron, 2004: 25; Ridsdale, 2007: 59). Fotherby negotiated with Caspian that club's chairman would still be local – himself in the first instance and then, the following year, Peter Ridsdale.

Ridsdale was, in some senses, local, having been born in the Hyde Park district of Leeds in 1952 and a Leeds United supporter of longstanding who'd stood for many years in the West Stand Paddock at Elland Road. But he was a career businessman who'd worked for the most part away from Yorkshire – successively for ICL, Scholl footwear, Burtons/Top Man and the QVC tele-shopping channel. He came back to Leeds in the 1990s as chief executive of Tulchan, a financial communications consultancy that was part of the 3i private equity group (Ridsdale 2007). He had come to the Leeds boardroom originally through Top Man's sponsorship of Wilkinson's Championship-winning side in 1991-2.

Ridsdale was, expectably for a man with his biography, an enthusiastic proponent of the new big bucks football culture that had followed the establishment of the Premier League: 'As much as supporters talk about greed and obscene money in football, they still cheer and chant these immensely well-paid heroes onto the pitch, applauding their celebrity contribution to the dreams that surround each club. I don't see unions of supporters waving placards outside the stadiums, protesting about capitalism' (Ridsdale 2007: 212-13). Indeed, the word 'dream' appears often in Ridsdale's memoirs, published in 2007; he shared O' Leary's dream for the club and agreed to fund his pursuit of it (Ridsdale 2007: 86-7).

In the limited space I have available, I'd like to consider two key aspects of the Ridsdale/O'Leary period: the arrest and trial of two Leeds United footballers in 2000-1 and the financial collapse of the club in 2003. Much has already been written on these two matters. I concentrate here on what these events seem to reveal about football and culture in the postmodern northern city.

Jonathan Woodgate was born in Revie's and Clough's Middlesbrough in 1980 and signed for Leeds at 18. Lee Bowyer was born three years earlier in Canning Town, East London and came to Leeds from Charlton Athletic in 1996. Their arrest in January 2000 and subsequent trial seemed to be Don Revie's most vivid nightmare come to pass.

Both young men had a record of violence. At the time he came to Leeds, Bowyer was already due in court after two Asian waiters were attacked in a McDonalds restaurant in London's Isle of Dogs. Bowyer later admitted affray in relation to this incident. Woodgate had been cautioned for assault at the age of 14 and arrested two years later following an attack on a student in Middlesbrough (*Guardian* 14th December 2001) He was banned from the bar of the University of Teesside and was said, when out drinking, to take notes from his wallet and set fire to them (Bradshaw 2001).

On the night of 11th January 2000, Woodgate was in Leeds city centre with four male friends from Middlesbrough and several Leeds United reserve players. Leeds city centre is, as is made clear elsewhere in this book, the epitome of the transformed, postmodern northern town. By day it is the location of the fabled 'Harvey Nicks' but by night, in keeping with the culture of the '24 hour city', other forms of consumption take over. These forms reflect a new politics of leisure and, in Leeds as elsewhere, various appetites – for food, drink, sex or drugs – are readily supplied in the bars, restaurants and clubs that have come to characterise the middle of the city since the 1980s. Thus, while their predecessors in the 1960s would probably have been home in their digs or playing carpet bowls with Don Revie, Woodgate's party was on a crawl that involved pint-glass rum and vodka cocktails in several bars and a visit to the DV8 lap-dancing club before coming to rest at the Majestyk club on City Square. Here they occupied the 'VIP lounge'. They were joined at some stage by Bowyer. In the Majestyk they encountered Sarfraz Najeib, a young student at Leeds Metropolitan University, his brother Shazhad and their three friends. It was later claimed in court by one of the club's bouncers that these five young men, all Muslims, teased the Woodgate party, by now very drunk, for not being able to hold their liquor. Sarfaz Najeib was later beaten unconscious in nearby Boar Lane (Rostron 2004: 30-1).

After one aborted trial, Woodgate was found guilty of affray and one of his Middlesbrough mates, convicted both of affray and of causing grievous bodily harm, went to prison for six years. Bowyer was acquitted.

In the age of global communication and an equally global preoccupation with the game of football, this case represented a serious threat to Leeds United as a brand and to manager O'Leary's expressed wish both to make the club attractive to television companies and to qualify for European competition. It also had commercial, social and political ramifications closer to home. Central to this was the spectre of 'race'. An apparent assault by drunken young white males on a teetotal Muslim youth clearly had an *a priori* suggestion of racism. The Majestyk, according to *Guardian* writer Gary Younge, had 'spent more than a decade struggling to shake off a reputation as one of the most racist in Britain' (Younge 2001). Moreover, in his original statement to police, the victim had said that one of his attackers had called out 'Do you want some, Paki?' He had, understandably, been unable to specify which assailant had said this and there were no independent witnesses to the remark. Subsequently the Crown Prosecution Service had dropped the racial aspect of the charges altogether (Younge 2001).

As Younge points out, a racial assault would have had more serious legal consequences. It would also very likely undo much of the community work Leeds United had been doing with ethnic minorities in the city in its efforts to combat racism: in the 1980s the fascist National Front had openly recruited at Elland Road. In any event, racial attacks in the city had increased thirteen-fold since 1995 (Younge 2001). Even acquittal or lenient sentences (a combination of which is what the players received) would not solve the club's difficulty here. On the one hand, the two men were expensive pieces of human machinery, paid thousands of pounds a week, worth millions in the football transfer market and key elements in the club's quest for European success. On the other hand, if the club did nothing, Leeds United would be considerably soiled as a brand, deterring sponsors and diminishing their chances of building an international 'fan base'. Ridsdale's business interests were also under threat: he chaired Education Leeds, a 'public-private partnership' which, availing itself of the 'New' Labour government's education policy, had taken over the city's schools in the wake of a damning Ofsted report (Younge 2001, Wainwright 2001). Ridsdale's failure as United chairman to condemn binge drinking and violent disorder would have called his company's stewardship of the city's young into question.

Leeds United responded with heavy fines for both players. A book called *Leeds United on Trial* also appeared. Credited to O'Leary (O'Leary 2002), the book is nevertheless believed to have been written by the club's communications director (*Guardian* Leader 18th December 2001.[4] The book is written out of a keen awareness of the Bowyer-Woodgate incident in relation to globalisation: 'It seemed that wherever I went in the world, from Madrid to Dubai to Rome, television networks such as CNN were providing daily bulletins on the fortunes of the Leeds United players at Hull Crown Court' (O'Leary 2002: 106). The book sought, to the dismay of some that read it, to turn the initial suggestion of racist intent back on the accusers, claiming that it sprang from ' a misplaced idea of political correctness' and 'a political undercurrent'. It had been a 'slur', withdrawn only when 'the damage had been done' (p.21). There was no racism at Leeds United, it asserted, because the club masseur was black (massaging Bowyer and Woodgate

---

4   http://www.guardian.co.uk/news/2001/dec/18/leadersandreply.mainsection Access: 3rd June 2008.

'on a daily basis'). Moreover, a member of the reserve team was ' a Leeds-born Asian' who could often be seen 'chatting and laughing with Bow and Woody'. Besides, O'Leary's parents, Irish migrants to Britain in the 1950s, had known racism too and he himself had suffered 'Paddy' taunts (p.22). The club, it was claimed, had received racist hate mail from 'misguided Asians [who] apparently wanted to declare a *jihad*, a holy war, on the players ....' (p.27). A television news bulletin, which referred carelessly to 'The Leeds United trial ...', was accused of inflicting 'an unnecessary smear on the vast majority of employees' at the club (p.93). While the book condemns racism, violence and excess drinking at several points, the strong implication for much of its 177 pages is that Leeds United, rather than Sarfraz Najeib, were the principal victims of the whole episode. Given this, the book maintains, the performances of Bowyer for Leeds during the trial were 'phenomenal' – in one game, against the Belgian club Anderlecht, there were 'deafening chants of "Bowyer for England"'. With the verdict not yet in, for many people concerned about racial harmony in Leeds, the Crown Prosecution Service's decision notwithstanding, this was likely to have been disturbing.

Despite the deployment of this formidable 'PR shield' (Ridsdale 2007: 145), it's difficult to think that Leeds United, the brand, did not suffer in the market place through these events. That said, it's doubtful if any damage done here was decisive in Leeds United's financial fall from grace in 2003. The club, of course, had changed a good deal since Revie's time. He had advocated more sponsorship in football (he negotiated shirt sponsorship for the England team in 1974) and there was now plenty of that: when Ridsdale was confirmed as chairman in June of 1997, it was in a banqueting suite in front of 300 corporate sponsors (Ridsdale 2007: 64). Sponsorship constituted vital income for a club like Leeds. By the late 1990s they were thriving on several fronts. Their turnover at the end of 2000 was £41 million, with their various 'revenue streams' – gate receipts, television fees, merchandising and commerce – all showing a clear profit (Rostron 2004: 40). This, by 1960s standards, was very big money and, as I've noted, the typical Leeds United director was no longer a local businessman, pursuing his hobby: Ridsdale, almost a cartoon version of the new breed of director, was reputed to receive a salary of £450,000 (Fifield 2003).

At Leeds, from the mid-1990s, there is a perceptible trend away from the local and toward men who painted on a broader corporate canvas: there were at various times chief executives, chief operating officers, specialist directors for finance, commerce, operations and so on. People came to the board through their (probably temporary) proximity to Leeds and their corporate experience. Allan Leighton, for example, was asked to become deputy chairman in 1998, when he was based in Leeds as chairman of ASDA, the supermarket chain. He lived mostly in Buckinghamshire and held eight other company directorships. Asked what he thought in retrospect of the dismissal of manager O' Leary (in 2002), he replied: 'It's always a difficult decision firing anybody. I'm just in the process of making 30,000 people redundant in the Post Office' (Rostron 2004: 127-9).

Ridsdale and O'Leary had adopted the strategy advocated at the beginning of this chapter by Professor Stefan Szymanski: they borrowed heavily against future success. In practice, their lucrative revenue streams had to be balanced against the huge costs of players transfer fees and salaries. The club took on huge loans to bring in players, gambling on qualification for European competition, – the deal to sign Rio Ferdinand from West Ham United in 2000, for example, was to be financed through hoped-for entry into the Champions League (Rostron, 2004: 139). This strategy failed and in 2003 the club was '£78 million in debt despite £52 million worth of players being offloaded' (Fifield 2003).

In the wake of this collapse stories of the spendthrift nature of Ridsdale regime became legion, the most favoured being the revelation that £20 per month was set aside for the renting and maintenance of the club goldfish (Fifield 2001).

Two things seem important to note in the wake of the failure of the Ridsdale/ O'Leary gamble.

One is the increasing prominence in the football world of experts in the commercial sub-discipline of financial salvage. Professor John Mackenzie, a Yorkshire-based economist, briefly succeeded Ridsdale as Leeds United chairman. He was followed by Trevor Birch (2003-4), a specialist in guiding football clubs out of financial turmoil and Gerald Krasner (2004-5), who worked for Bartfields, a corporate recovery firm, and had been a member of the Department of Trade and Industry team brought in to investigate the club.

The other is the emergence of football club chairmen as celebrities. This is due in part to the greater commercialisation of football clubs and in part to the more central place occupied by entrepreneurs in popular culture: BBC television's *The Apprentice* (2005-), in which aspiring young people compete for a job under the self-consciously brutish Sir Alan Sugar (computer magnate and chairman of Tottenham Hotspur, 1991-2000) is an example of this. The current chairman of Leeds United is Ken Bates, a successful businessman in the global arena (haulage, quarrying, readymix concrete, dairy farming ....), outspoken, high-profile and a proponent of the same uncompromising business values as Sugar and Ridsdale. These are merely a few of the large number of men now interested simply in owning and/or running football clubs. They do not, as in Revie's time, aspire simply to act as stewards of their local one – partly because they are unlikely to acknowledge the concept of 'local'. Bates has been financially involved with Oldham Athletic, Wigan and Chelsea as well as Leeds and lives in Monaco. The Serbian-American tycoon Milan Mandaric has owned San Jose Earthquakes (in the USA), Standard Liege (in Belgium), Portsmouth and Leicester City and resides largely in the United States. Ridsdale has owned Barnsley and Cardiff City in the comparatively short time since he left Leeds United.

The recent history of Leeds United illustrates a wider truth of contemporary football club culture at the elite level: publicity is both its principal commodity (what it offers sponsors) and its greatest intoxicant: Professor Mackenzie attended meetings of Premier League chairmen where, he said, there were 'a great many egos'. Ridsdale, his predecessor, known as 'Publicity Pete' to the press corps

(Rostron 2004: 111) was 'heavily addicted to the business and that visibility issue' (Rostron 2004: 166). That is, Ridsdale, among other things sought fame – a notion not dented by the publication of his autobiography in 2007 (Ridsdale 2007). He wanted to be a football director celebrity – some boardroom approximation, in effect, to Brian Clough.

The Blessed United: Revie's Team in the Theatre of Memory

The historian Raphael Samuel appears to have coined the term 'theatres of memory' in the early 1990s. In the first of his two-volume treatise on the uses of history, which bears this title (Samuel, 1994: 1998), Samuel defended aspects of the burgeoning 'heritage industry' against its critics, rejecting what he saw as 'top-down accounts' of the apparent growth of museums and the commodification of the past (Samuel 1994: 242-56). Unlike writers such as Patrick Wright, who saw heritage as the reactionary celebration of aristocratic values (Wright 1985) and Robert Hewison, who denounced it as a 'denial of the future' (Hewison 1987: 46), Samuel recognised heritage in a far wider range of activities than simply the opening of a country house to the public or the founding of a museum. And, unlike its detractors, he argued that it could express a vibrant and democratic relationship to history, seeing it as 'an *escape* from class. Instead of heredity, it offers a sense of place ....' (Samuel 1994: 246).

This preparedness to assess memory and heritage in a more rounded way – encompassing, say, steam engine rallies, charity shops, flea markets .... – can be bracketed with a parallel trend in cultural judgment. It is an axiom of postmodern culture that distinctions between high and low culture are sundered (Strinati 2004) and this is seen to be reflected in recent trends in heritage. Thus, whereas 'heritage' ordinarily memorialised figures from high culture (poets, novelists, composers) or affairs of state (royalty, politicians, soldiers ...), today it is just as likely to celebrate people from popular culture: a blue plaque may, for example, go up on the wall of a flat once occupied by Jimi Hendrix, a doodle by John Lennon may fetch thousands of pounds at auction, and so on. Football, similarly, has its own heritage industry: a museum at Preston (opened in 2001), a flourishing after-dinner speaker trade, DVDs of famous matches and an apparently huge growth in football literature, encompassing footballer biographies, club histories, diaries of a season and the like.

In contemporary, postmodern Leeds, Don Revie and his players live on – some in person, all collectively as myth. 'They were great days, heady days now in the dim and distant past', wrote Norman Hunter in 2004, 'but they remain memorable for me and for the supporters who followed the team at the time, as I have found out on my travels around the country as an after dinner speaker' (Hunter and Warters 2004: 25). In a book published in 2005, detailing the whereabouts of past Leeds players, Paul Reaney (Leeds United 1961-78) is said since 2000 to have been running a business called 'Golden Era'. Reaney 'has pooled together all the old Revie lads and coordinates their work and appearances. They're in great demand.

There are 20 players and he does most of their commercial deals, from supermarket openings to after-dinners and books' (Rowley, Wray and Endeacott 2005: 118). And, of the personnel of the Revie era at least a dozen players – Gary Sprake, Bobby Collins, Johnny Giles, Jack Charlton, Billy Bremner (d. 1997), Norman Hunter, Terry Yorath, Paul Madeley, Peter Lorimer, Allan Clarke, Mick Jones and Eddie Gray – as well as Revie himself, have been the subject of biographies. Four of these (on Collins, Madeley, Clarke, Jones), along with, at the time of writing, ten other Leeds United books, were written by local writer David Saffer. The books are consciously aimed at the nostalgia market: the introduction to each ends with the words 'Enjoy the memories' (Saffer 2001; 2002; 2003; 2004).

Saffer began writing in 1997 when he called Leeds United and learned that the club had no plans to mark the 25th anniversary of the Revie team's F.A.Cup win in 1972: 'Ten years ago there was no corporate hospitality, or corporate events or after dinner speaking. It was in the very early days of that type of thing – when I started out writing, it was very easy to have access to the older players. Ten years down the line a number of the players I deal with are now on the after dinner circuit.' In the late 1990s Saffer proposed to one of the Revie team 'a *Boys Own*' type biography, where there's lots of pictures and memorabilia and all your memories floated in so you've got pictures throughout and your story around it, so it's more like a scrapbook … it sold thousands … the four biographies I've written – Clarke, Jones, Madeley and Collins – it's very much a local market, a Leeds United market. They're sold in all the shops around here. But they're sold all over the country. There are Leeds fans everywhere'. 'The Revie team', he suggests, 'had a massive effect …' 'A number of them do corporate hospitality every home game now…' A number of re-union 'do's' for the Revie players have, he says, been 'absolutely packed out … hundreds of fans …'[5]

What sense can be made of this lingering presence in Leeds, and among Leeds United supporters scattered far and wide, of the ghost of Don Revie? In addressing this question I draw on comments on the Revie era that I invited from members of the Leeds United Supporters' Trust by email in June 2008.

Since the 1960s football and 'their' football club has become a more important source of identity for many. One of my respondents said: 'My father's generation took a passing interest and if they attended games they did so more as spectators rather than as fans. My own generation is different. I cannot remember a time without football and Leeds United'. In a recent talk [6], the journalist Anthony Clavane discussed the subjectivity – his and others' – that is labelled 'We are Leeds': 'We are Leeds and we do heroic defiance, but we also do unheroic failure … it seems to me that we are at our best when we do heroic defiance, when everyone hates us and we don't care, but at our worst when – having jumped through every hoop, overcome every obstacle – we reach a final, a crucial decider, the last stage

---

5   All quotations from interview with the author, Leeds 1st May 2008.
6   'Hubris, Schmubris …'

of a heroically defiant campaign. What Everest climbers call the Hillary Step. It is then that we fall back down the slippery slopes of failure'.

Don Revie and his team embodied this sense of self – as we saw earlier, before him Leeds United had only 'prehistory'. Revie, for all the respondents, was the object of admiration: 'a true legend of the city', 'revered in Leeds', 'our best ever manager and a true gent', 'the man who put Leeds United on the map' and so on. One judgment was interestingly qualified: 'Revie was a great man, but slightly neurotic and too cautious. [The T]eam could be cynical but played some superb football' and another dubbed Revie 'brilliant, unlucky, always the bridesmaid ...'. Revie's side, for one, was remembered 'mainly as heroes and still talked about... we were discussing the 68/69 championship side in the pub last night!' It was generally acknowledged, however, that this opinion might not stretch far beyond the West Riding: one said: 'I think Don Revie is remembered with great affection in Leeds, certainly by those who recall that period. This is not the case in the wider world or by those outside the 'Leeds Utd Family'. In my opinion Leeds United and Don Revie in particular have been demonised from the mid sixties right up to the present day'. Another respondent thought the outside world remembered Revie as 'a "nearly man" who let down England and whose team won by cheating, fouling and bribing'. 'Plenty of fans were jealous' explained a third.

Having in 1961 changed the team's colours to all-white, in honour of Real Madrid Revie saw the day when his team, in beating Southampton 7-0 in March of 1972, drew cheers of 'Ole' from the Elland Road crowd for the imperious quality of their football (Bagchi and Rogerson, 2002: 150-2). The youngest of my respondents – a 26 year-old civil engineer living in Bramley, West Leeds – expressed the 'absolute joy' of watching this game on *YouTube* but there was general acknowledgement that the Revie players are nevertheless remembered as 'Dirty Leeds'. This was largely attributed to the depictions of Leeds in the national media: 'They couldn't wait to see us fail', said one respondent.

A 49 year-old Systems administrator from Wakefield wrote:

> Many of the Leeds fans born within a few years either side of my own birth date are amongst the most passionate and loyal supporters in football. They have a deep emotional attachment to the club. Without the Leeds Utd of the Revie era these fans would not exist.

These various responses were, in large part, expectable but it is still important to interpret them. Revie, doubtless with the assistance of the national football media, became a metaphor for the Sixties; in turn this most metaphorical of decades became synonymous with football's purported degradations: commercialism, intimidation, cheating. The 59 year old merchandising manager who described Revie as a 'great tactician' who produced 'the first truly "professional" team' was expressing a popular view. Growing professionalism in the game was, of course, a social phenomenon; in popular discourse, however, it became the work of one flawed individual and his fractious protégés.

This 'flawed individual' was, of course, from the North of England. Here, historically, there has been much resentment of 'the South': the mythical[7] locus of power, privilege and condescension. There was a residual notion that the social dice were loaded against northerners – Yorkshire cricketers of the mid-twentieth century, for example, were wont to think they had to be twice as good as southern ones to get into the England side – and this formed part of their rationale for any failure to gain selection. The Revie myth is plainly marriageable with longer standing and wider ranging myths of the north and, in particular, with typical 'We are Leeds' subjectivities. Here, the thought might run, was a man who taught a football nation of amateurs how to play the game successfully and was repaid with scorn. Revie sensed he wouldn't get a fair crack of the whip from reporters, referees or administrators so, where he could, he tried to order the universe, and, where he could not, like so many of his admiring public, he resorted to superstition – his lucky suits and the rest.

Revie made the local global: he took Leeds United into Europe and, as a consequence, made many of the people of a northern town feel good about themselves. The 'Leeds United family', whether or not they were around at the time, remember him for this and they also remember the anger he inspired in other quarters; this in turn, as Clavane observes, becomes part of their identity – 'You all hate us ....'. In this regard, *Leeds United on Trial* is, arguably, a corporate and sophisticated draught of revived Revie-ism, with the club now menaced, not by conniving F.A. officials and incompetent ref's, but by 'political correctness'.

Aside from this sense of *place* (Leeds, the North…) the Revie team in the theatre of memory also takes people back to a specific *time* – when footballers lived in the town and not in expensive commuter villages ('Paul Madeley used to get to the ground on the bus', said David Saffer) and when the city still had a working class and when it produced *things* (woollens, tanks, board games…) and not just the services (the shopping, the latte coffees, the office space …) that now define the postmodern city. 'A few weeks ago', said a Leeds supporter at a recent public gathering, 'there was an article in the *Yorkshire Evening Post* saying that 30% of households in Leeds now contain no employed person'. 'Yes', said another, 'people talk about the shiny new Leeds of shops and offices and city centre flats. Nobody mentions all the estates'.[8]

Acknowledgments

The film *How Green Was My Valley* was scripted by Philip Dunne, directed by John Ford and released by Twentieth Century Fox in 1941. It was based on the

---

7   Importantly I'm using the word 'mythical' in the way suggested by Roland Barthes and employed to great effect by historians such as Angus Calder. Barthes took the word to mean 'depoliticised speech' – see his *Mythologies* (St. Albans: Paladin 1973 pp.142-5). Myths, therefore, were understood be notions that were, in effect, 'naturalised' and could never be discussed or challenged. See, for example, Calders' *The Myth of the Blitz* (London: Pimlico, 1992).

8   'Hubris, Schmubris …' – contributions from the floor.

novel by Richard Llewelyn. Either Dunne or Llewelyn could have originated the passage which prefaces this chapter. It is reproduced in Joseph McBride *Searching for John Ford: A Life* (London: Faber & Faber, 2003, pp. 41-2).

Herb Behrens, of the National Steinback Centre in Salinas, California was kind enough to trace the quotation from *Cannery Row*, which had been in my mind for many years.

Many thanks for their help in the preparation of this chapter to Rob Bagchi, John Boyd, Pete Bramham, Anthony Clavane, Tony Collins, Richard Dawson, Rick Duniec, David Jackson, Dave Russell, David Saffer, Rob Steen and Beccy Watson.

References

Bagchi, R. and Rogerson, P. (2002). *The Unforgiven: The Story of Don Revie's Leeds United.* London: Aurum Press.

BBC Sport. (2008). Revie Shirt Taken in Leeds Theft. [Online, 4 June] Available at http://news.bbc.co.uk/sport1/hi/football/teams/l/leeds_united/7433396.stm [accessed: 4 June 2008].

Bradshaw, B. (2001). The Rise and Fall of Woodgate. *The Observer* [Online,16th December] Available at: http://www.guardian.co.uk/football/2001/dec/16/sport. comment2 [accessed: 3 June 2008].

Bremner, B. (1969). *You Get Nowt For Being Second.* London: Souvenir Press.

Chapman, H. [1934] (2007). *Herbert Chapman on Football.* Robert Blatchford [First published 1934, London: Garrick].

Charlton, J. with Byrne, P. (1996). *Jack Charlton: The Autobiography.* London: Partridge Press.

Clough, B. with Sadler, J. (2005). *Cloughie: Walking on Water.* London: Headline.

Corbett, J. (2008). The King of the Damned. *Observer Sport Monthly* [Online, 25 November] Available at http://www.guardian.co.uk/sport/2007/nov/25/football. newsstory [accessed: 1 May 2008].

Doyle, P. (2005). Small talk: John Giles. *The Guardian* [Online, 29 April]. Available at http://www.guardian.co.uk/sport/2005/apr/29/smalltalk.sportinterviews [accesssed: 1 July 2008].

Fifield, D. (2003). New Leeds Chairman Uncovers Fishy Past. *The Guardian.* [Online, 21 May] Available at: http://www.guardian.co.uk/football/2003/may/21/ sport.comment1 [accessed: 1 May 2008].

Goffman, E. (1991). *Asylums.* Harmondsworth: Penguin.

Greaves, J. (2003). *Greavsie: The Autobiography.* London: Time Warner Books.

Guardian. (2001). Woodgate and Bowyer had Violent Histories. *The Guardian [Online* 14 December] Available at http://www.guardian.co.uk/football/2001/ dec/14/newsstory.sport13_[accessed: 3 June 2008].

Hewison, R. (1987). *The Heritage Industry.* London: Methuen.

Hopcraft, A. (1968). *The Football Man.* London: Collins.

Hunter, N. with Warters, D. 2004. *Biting Talk: My Autobiography.* London: Hodder and Stoughton.

Jordan, J. with Lawton, J. (2004). *Behind the Dream: My Autobiography.* London: Hodder and Stoughton.

Moynihan, L. (2007). *Gordon Strachan: The Biography* London: Virgin Books.

O' Leary, D. (2002). *Leeds United on Trial* London: Little, Brown.

Page, S. (2006). *Herbert Chapman: The First Great Manager.* Birmingham: Heroes Publishing.

Peace, D. (2006). *The Damned United.* London: Faber and Faber.

Ridsdale, P. (2007). *United We Fall: Boardroom Truths About the Beautiful Game.* London: Macmillan.

Risoli, M. (2003). *John Charles: Gentle Giant.* Edinburgh: Mainstream.

Rostron, P. (2004). *Leeds United: Trials and Tribulations.* Edinburgh: Mainstream.

Rowley, L., Wray, J. and Endeacott, R. (2005). *Where Are They Now? Life After Leeds United.* Leeds: YFP Publishing.

Russell, D. (2004). *Looking North: Northern England and the National Imagination.* Manchester: Manchester University Press.

Saffer, D. (2001). *Sniffer: The Life and Times of Allan Clarke.* Stroud: Tempus.

Saffer, D. (2002). *The Life and Times of Mick Jones.* Stroud: Tempus.

Saffer, D. (2003). *Leeds United's 'Rolls Royce': The Paul Madeley Story.* Stroud: Tempus.

Saffer, D. (2004). *Bobby Collins: Scotland's Mighty Atom.* Stroud: Tempus.

Samuel, R. (1994). *Theatres of Memory. Volume 1: Past and Present in Contemporary Culture.* London: Verso.

Samuel, R. (1998). *Island Stories: Unravelling Britain. Theatres of Memory, Volume II* London: Verso.

Say, T. (1996). Herbert Chapman: Football Revolutionary. *The Sports Historian,* 16: May, 81–98.

Sillitoe, A. (1960). *Saturday Night and Sunday Morning.* London: Pan Books.

Smith, G. (2006). Hated? Yes. Boring? Maybe. Today's equivalent of Revie's Leeds? Never. *The Times* [Online, 4 December] Available at http://www.timesonline.co.uk/article/0,,8304-2124068,00.html [accessed 4 December 2006].

Strinati, Dominic (2004). *An Introduction to Theories of Popular Culture.* London: Routledge.

Studd, S. (1998). *Herbert Chapman, Football Emperor.* London: Peter Owen.

Wagg, S. (1984). *The Football World: A Contemporary Social History.* Brighton: Harvester.

Wainwright, M. (2001). Fans Stand by their Man. *The Guardian* [Online, 19 December] Available at http://www.guardian.co.uk/football/2001/dec/19/newsstory.sport3 [accessed: 3 June 2006].

Wright, P. (1985). *On Living in an Old Country.* London: Verso.

Yorath, T. with Lloyd, G. (2004). *Hard Man, Hard Knocks.* Cardiff: Celluloid.

Younge, G. (2001). Letting the Side Down. *The Guardian* [Online, 17 December] Available at http://www.guardian.co.uk/world/2001/dec/17/race.football [accessed 3 June 2008].

Chapter 8

Dreams of Parkside and Barley Mow

Karl Spracklen

In the dim light of the inferior arc lamps around the pitch I could just make out, through the mist, dark and ferocious giants running and grunting and passing between them a grey ball. My hands, warmed by a portion of chips, were still too cold to clap, so I cheered with every cheer that went up. I knew who I was supporting: it was my dad's team, our team, the one whose scarf and bobble hat adorned my head and body. New Hunslet Rugby League Football Club. I wasn't entirely sure why we were New, but I did know Hunslet was an old club, once the biggest and best club in Leeds. I knew my dad had taken me to Parkside, the ground Hunslet had played at for most of their existence. Hunslet was my dad's club. "We've swept the seas before, boys, and so we shall again", was our song. Green and white were our colours (actually, the original colours were myrtle, white and flame, but that combination had yet to evoke any feeling of loss inside my child's mind).

I watched the giants grapple with each other in the mist, grunting and heaving. I could see players catch the ball and attack the try-line. But the game itself was beyond the understanding of a five year old. We were playing our rivals from the other side of town, Leeds, the old enemy, the team my dad cursed and swore at whenever they were mentioned in the paper. So I, naturally, felt the same way about them. Yet I cheered with every cheer, unaware that most of the cheers were for the other team, until my older brother punched me.

Then, from out of the mist, a player leaner and smaller than the others appeared, running right past where I stood with the ball in his hands. I looked at him and recognised him as one of my players: Rudy Francis. He had a number five on his back, and his jersey was clean enough to be distinguishable from the others, but it was his face I knew. He was my hero, he stood out from the rest of the players, who were all granite faced and sported fashionable mutton-chop sideburns. He was a player even a five year old, with limited attention spans, could recognise.

He was black, but at the time I didn't understand what that meant. All I knew was he scored tries, and I always wanted to be him whenever we played. I remember seeing the sweat on his face, the spittle at the edges of his mouth, and the draught of wintergreen and damp grass as he raced by. I remember, though I was too small to see, the cry that went up from my dad and brother as he touched down unopposed. I couldn't understand the jubilation on the faces of the adults who were stood around me, but I cheered and shouted his name anyway. Then my team's name was chanted, and I imagined that I was my hero, waving at my fans, and I felt I belonged there. When the hooter went my dad told me we had won – and what's more, we had beaten *them*: beaten Leeds, the fancy folk from Headingley.

Rugby league's heartlands are small towns of the north of England. Before the collapse of the country's heavy industries, rugby league clubs had strong associations with the work of their supporters. Many amateur rugby league clubs relied on local factories for financial support and sports facilities. Professional rugby league clubs were sponsored by local factories, and players found employment there. A large number of professional rugby league clubs were to be found in mining areas: Wigan, Saint Helens and Leigh in south-west Lancashire; Workington and Whitehaven on the Cumberland coast; and the West Riding of Yorkshire's Castleford, Featherstone and Wakefield. Even in Leeds, there were mines on the south side of the city, and the last one survived until the end of the Miners' Strike in 1985. Other clubs were in towns associated with other heavy industries: Widnes owed its existence to the ICI Chemical Works that gave the club their nickname; Hunslet was in a part of Leeds famous for its railway engineering; Dewsbury and Batley were part of an area still known as Heavy Woollen, dominated by textiles.

Reading the rugby league results was like reading a concise history of the industrial north, from fishing fleets on Hull to wire works of Warrington. Rugby league was part of the north, part of its industrial landscape: and, when the industries declined at the beginning of the 1970s, rugby league also fell into a spiral of falling crowds and failing clubs. Amateur rugby league almost disappeared completely at that time as factory teams were lost with the factories and men had little money to spend in club bars or on the terraces.

In 1984 Margaret Thatcher took on the National Union of Mineworkers and the rugby league belt of the north of England (sometimes called the M62 belt after the motorway that conveniently links up most of the northern league-playing towns) was galvanised by the prospect of class war. Club owners, new-money types with a tendency towards Freemasonry and golfing trips to the Algarve, were held under close scrutiny by the fans. Some club directors lent their support to the striking miners because they remembered their roots; others, wisely, kept their silence but didn't intervene to stop collections for the Strike Fund. Many of rugby league's elite players were miners themselves, or ex-miners, or the sons of miners. The rugby league community was on the side of the workers. Arthur Scargill was a hero.[1]

On the west side of Leeds where I grew up, there hadn't been any mines since the end of the nineteenth century. But we felt solidarity with the miners and workers who would lose their jobs servicing the mining industry if the industry collapsed. In school, the one boy who dared to defend Margaret Thatcher was ridiculed by the rest of us and called a posh traitor. At home, my mum had enormous rows with

---

1   A hero to the fans and probably those players with links to the coal mines: it can be assumed that the average First Division professional, and most club directors, leaned with their bags of money towards Margaret Thatcher.

my dad about gradualism or activism in the left-wing struggle against capitalism. My dad thought the strike was doomed to fail; my mum thought this amounted to treason against the cause. I was twelve years old but I was already familiar with arguments put forward by various shades of socialism. My mum came from Swinton in South Yorkshire: a mining area, but not a rugby league one (not to be confused with Swinton near Salford, which was rugby league territory). My dad didn't consider South Yorkshire to be part of the north: he claimed their accents were more like those in Nottinghamshire, which was deemed part of the Midlands. That kind of talk really annoyed my mum because the miners in the Midlands were all scabs. My mum's grand-dad had been a miner most of his life and died of lung disease; her dad, my grand-dad, escaped the dust and heat of the mine because of WW2 and the demand for brick-layers after it. But even if her dad had been a brickie instead of a pitman she still came from a mining community. My dad came from a rough part of Leeds, from a poor family, but he didn't have a *bona fide* link to the struggle. So my mum usually won the arguments: even now, when they're rowing, she'll berate my dad for criticising Scargill.

A few years later, when the strike was lost and most of the pits were closed, the Rugby Football League naively accepted sponsorship from British Coal, the successor to the National Coal Board that had done Margaret Thatcher's bidding. It's a mark of the sense of solidarity and pride amongst the northern working-class that at the 1992 World Cup Final between Great Britain and Australia, held at a packed Wembley Stadium, thousands of fans wearing Great Britain shirts defaced the British Coal logo emblazoned on the front of them.

The sport of rugby league, like all modern sports, has changed over the last thirty years due to the pressures of commodification (Hughson and Free, 2006), globalisation (Maguire, 2005) and the transition to post-industrial and postmodern identity formations (Wheaton, 2004). Its historical development as a variant of rugby union played in the north of England limited its ability to expand internationally for much of its existence (Collins, 2006).[2] Once a sport limited to working-class communities in England, Australia, New Zealand and France, rugby league is now played in over thirty countries, and every year the game's emic literature reports on new horizons and new ventures ranging from Argentina to Georgia. Since the switch from winter to summer rugby in the UK following the development of Super League and the involvement of Sky Television (Denham, 2004; Collins, 2006), rugby league has started to establish a national supporter base and national focus. Commercialisation and commodification of rugby league in this period of late modernity are typified by the business model of Leeds Rugby Club and its partnership with Leeds Metropolitan University. With a string of

---

2   The popular debate about the "Split" in 1895 of the Northern Football Union from the Rugby Football Union is polarised by supporters of both codes into a story of righteous resistance versus a story of 'Johnny-come-lately' greed (Spracklen, 1996b). Even academic accounts offered by Collins (2006) are questionable in relation to the author's own rugby-league supporting bias.

international players and thousands of supporters (in the stadium and around the world watching television), Leeds Rhinos are at the heart of the globalised thirteen-a-side code. However, despite the lure of Carnegie Headingley Stadium, two other professional rugby league clubs survive in the shadow of the new Leeds: Hunslet Hawks, the first club to win all major trophies in the Northern Union (now the Rugby Football League); and Bramley Buffaloes, who disappeared after merging with Leeds but reformed as a community-led club in National League Three. For players and supporters in the north of England, rugby league was and is associated with the culture and history of the working-class north: with what was identified in previous research as "northerness" (Spracklen, 1996b, 2001). As suggested earlier, the north of England associated with rugby league is a narrow geographical area, confined to certain towns in the pre-1974 counties of Lancashire (in the south of the old county), the West Riding of Yorkshire[3] (and the outlying cities of Hull and York), and the west coast of Cumberland. Places in the geographical north that fall outside the rugby league heartland are not, according to people involved in rugby league, to be seen as properly northern:

> [Rugby league] showed me a way of living, admitted me into a world where I belonged ... rugby league has this love affair with its people, its geography, you can't separate it from where it is, it is so involved. (Spracklen 1996b: 235)

The north is where rugby league is played: northerners are the people, the men who play and watch the game, who create the northerness that is held up as the archetype of what it means to be a white, working-class, northern man. This northerness is itself part of the way in which the imaginary community of 'the game' is constructed and sustained through a shared (imagined) history of belonging and reinvention of traditions. It is constituted in shared knowledge of stereotypes and symbolic boundaries of the north of England – the mills, the pits, the pints and pies – but also a reverence for the ideals of northern, working-class masculinity: players as hard as nails, men who take no nonsense, who are tough but fair, who are as gritty as the stone that divides the north in two along the ridge of the Pennines:

> It's like ... everything about it; it's always been about honesty, about being our game, pride in yer roots ... A northern game, for northern folk, that's what it's always been ... you see we went our own way because we wanted to rule our own destiny ... and the people who play it, they've always reflected their roots

---

    3   It is a matter of coincidence only that most of the towns and districts of the West Riding that did not play rugby league were removed from the political boundaries of the metropolitan county of West Yorkshire created in 1974. That the metropolitan county of West Yorkshire had as its administrative capital the city of Wakefield – a city without a professional football team but with three professional rugby-league clubs – must be another coincidence.

... you don't get any egos, well you didn't until they started all this contract business, I'm talking about back then, before Lindsay came along, players lived in the same street as the fans, drank in the same pubs, so they were like ... it were an accepted part of the game, that it was ours, and that we, that is the hard working class I suppose, your average man, could somehow have his values expressed through rugby [league]. (Spracklen 1996b: 223)

This chapter explores contradictions of rugby league as both a globalised sport and as an imaginary, imagined community associated with working-class Leeds of the past. In this comparison of the three clubs, and the game of rugby league, some answers to the question of the impact of postmodernity on identity formation will be rehearsed, and the importance of traditional rugby league as a symbolic reminder of the history of Leeds will be emphasised. Part of this chapter is based on an analysis of existing archive material published on current official web-sites of the three professional rugby league clubs in Leeds[4], as well as information on league tables from the Rugby Football League.[5] Research findings on the current situation of rugby league in Leeds and elsewhere, the history of the game and its relationship to class and identity, come from twelve semi-structured interviews with key supporters of rugby league in 'traditional' Leeds. These supporters are individuals known to the researcher for their work in re-establishing Bramley RLFC and volunteering in rugby league for Hunslet RLFC or other amateur clubs in the South Leeds area. All respondents signed informed consent forms, and pseudonyms will be used as a matter of ethical good practice. Further information about rugby league as a global, postmodern game has been gathered from twelve participants on an open access, on-line forum promoting the development and expansion of rugby league. These participants were approached through that forum and provided with a series of open-ended questions to answer via email.

The History of Rugby League in Leeds

In the first two decades of the twentieth century, the biggest, most popular version of football in Leeds was the version of rugby played by the Northern Union. Association football (soccer) had yet to take a firm hold in Leeds – no sports journalist of the time could ever have predicted the dominance of Leeds United in later years.[6] In the north-west suburbs of the rapidly expanding city were the 'Loiners', the Leeds club that then, as now, flaunted its wealth and attracted

---

4   www.leedsrugby.com, www.hunslethawksrl.co.uk, and www.bramley-rlfc.co.uk (all accessed 15 May 2008).

5   Published in the now defunct Rothman's Yearbook series (edition used: Fletcher and Howes 1995).

6   Leeds City FC had been formed in 1904 and was based at the old ground of Holbeck NUFC. They joined the Football League in 1905 and played in the Second Division, but

crowds from across the north of the city. Leeds had originally started out as Leeds St John's in 1870, but by 1890 they were Leeds Cricket, Football and Athletic Club – Leeds RFC – based at the new Headingley Stadium. South of the river, two more professional Northern Union clubs ensured the local press was always full of reports about the rugby played by the Northern Union.

Opposite the Headingley stronghold of Leeds, visible from the terracing of Headingley Stadium, the suburb of Bramley proudly reminded everyone of its independent past with its club, the Villagers. Bramley were considered to be the junior of the three professional Northern Union clubs. They had joined the Northern Union in 1896, and were perennial strugglers. In the first two decades of the twentieth century Bramley's Barley Mow ground became a place feared by visiting teams from Leeds and Hunslet. And despite only ever winning the BBC2 Floodlit Trophy in the 1970s (a competition that they won in daylight due to power shortages), Bramley had a small and loyal following of working-class fans. When Bramley moved their ground from Barley Mow to Maclaren Field the move was literally over a wall; the Barley Mow pub remained adjacent to grounds old and new. Bramley's support remained respectable for a small semi-professional rugby league club into the 1980s but the changing demographics of the old village, and a series of financial problems affected success on the pitch, resulted in a sharp decline during the 1990s. These problems were compounded by an estimate for the cost of safety work at the ground that Bramley could not afford. The sale of Maclaren Field to become a private housing estate in 1995 saw Bramley move out of the village and into the Aire Valley, where they played two seasons at Headingley RUFC's Clarence Field. The move saw many supporters give up on the club. When in 1997 the club moved to share the ground of Leeds at Headingley Stadium, with financial and controlling links created between the Super League club and the smaller club, even more supporters gave up on Bramley. At the end of the 1999 season the club, if it could still be called a club, resigned from the Rugby Football League. The management of Leeds Rhinos floated the idea of keeping the name, at least to brand a feeder team, but the Bramley fans set-up a co-operative to bring the club back into existence.[7] The supporters' co-operative applied to the Rugby Football League and were rejected twice but on the third attempt in 2005, they were accepted into the new National League Three. The club's new ground was at Stanningley, the suburb of Leeds just down Broad Lane from Bramley town centre, and they were re-branded as the Bramley Buffaloes. Over 2,000 people turned out for their first match, where the Leeds Co-op, the club's sponsors, handed out free bars of

---

never really troubled the rugby-dominance of the city. They suffered financial problems and folded in 1919. Leeds United was established following the demise of Leeds City.

7   Not just Bramley fans. Some Hunslet fans, including the author's brother and father, aggrieved by the refusal of the RFL to allow the club to enter the Super League in 2000, crossed the divide and helped the Bramley supporters create their co-operative club. The first Chair of the new club was my brother, Lee Spracklen.

Fair-Trade chocolate. The club has continued to perform well in the successor league to National League Three, and still attracts respectable crowds.

The second club prominent at the beginning of the twentieth century was much bigger than Bramley. A few miles to the east of Bramley and Headingley, the working-class industrial district of Hunslet, then and now a heartland of rugby, had its own club based at Parkside. From the earliest days of rugby football in Leeds, Hunslet had been one of the city's strongest and most-popular clubs. As such, the rivalry between Hunslet and its neighbours, Leeds, was long-standing. Hunslet's status as a senior rugby football club and the only senior rival to Leeds RFC in the city was secure by the 1890s. Hunslet, like Leeds, became one of the founding clubs of the Northern Union. Northern clubs run by middle-class entrepreneurs, with large working-class fan-bases, wanted rules on professionalism to be relaxed. These clubs wanted to end the hypocrisy of 'shamateurism' and allow players to be paid openly. Their supporters increasingly questioned why their game had to be run according to arcane rules about amateurs and professionals dictated to them by southerners. 'The Split' of 1895 led to the founding of the Northern Rugby Football Union, initially a league of clubs playing to the rules of the Rugby Football Union but allowing players to be paid for time off work (Collins 1999). However, the Northern Union soon changed the rules of rugby to make the game more attractive to paying spectators: line-outs were abandoned, two forwards dropped from play, and the ruck and maul ultimately replaced by the play-the-ball (Collins 2006). In the first season of the Northern Union, Hunslet finished in seventh place ahead of Leeds at number twelve in the league table. In the next year, when the Northern Union had expanded with the admission of other northern-based rugby clubs (necessitating a split into Yorkshire and Lancashire divisions), Hunslet came fourth in the Yorkshire Senior Competition and Leeds struggled behind in twelfth position again. Then, in 1897-1898, Hunslet won the Yorkshire Senior Competition, with Leeds down the table in ninth. When Leeds Parish Church NUFC and Holbeck NUFC both folded (in 1903 and 1904 respectively), Hunslet extended its supporter-base into much of the eastern parts of Leeds. In 1907-08, Hunslet Northern Union Football Club had an all-conquering season that saw them become the first professional rugby club to win all the trophies on offer.

Hunslet's dominance of the Northern Union in Leeds continued into the era of the Rugby Football League[8], despite a decade of failing to win any trophies in the 1920s. Although Leeds United's popularity grew with the rise of soccer in the inter-war years, rugby league continued to be central to the sporting lives of respectable working-class Leeds. Hunslet won the Challenge Cup in 1932, and in 1938 Hunslet beat Leeds in the Championship Final in front of a (then) record crowd of 54,112.[9] One of Hunslet's strengths lay in the close connection between the club and the working-class men who lived and worked in the district

---

8 The Northern Union renamed itself the Rugby Football League in 192x, adopting the Australian name rugby league for its version of the rugby code.

9 The match was played at Elland Road, the home of Leeds United.

(Hoggart 1957). The disruption of the Second World War, and the slow decline of Hunslet's industrial core, saw the club struggle in the 1950s, and although the brilliance of Hunslet's local-born players saw the club reach the Challenge Cup Final in 1965, this was to be the club's last appearance in any major final. The district of Hunslet saw most of its terrace streets knocked down and replaced by a new shopping centre and the Leek Street Flats, a housing complex inspired by the harsh modernism of Corbusier. Many Hunslet residents left the district to live on new estates built by Leeds City Council in Middleton and Morley. By the end of the 1960s the club was in financial difficulties: in 1971 arsonists burned down the main stand at Parkside; one year later the ground itself was sold off to become an industrial estate, and Hunslet RLFC disappeared altogether.

The club was resurrected as New Hunslet in 1973 by Geoff Gunney, one of the Wembley squad of 1965, and an ex-Great Britain international. Along with a few hundred supporters and sponsors, he lobbied for the new club to be accepted into the Rugby Football League, and despite some attempts to block the new club, it started the 1973-1974 season and re-established themselves. But with Parkside lost under warehouses, the new club was forced to play out of the Greyhound Stadium opposite Leeds United's Elland Road. The 1970s were hard years for Hunslet supporters. The Greyhound Stadium was unwelcoming, and not on an easy bus-route. The new club struggled to make any impact in the new Second Division for a number of seasons. The district of Hunslet's demographic profile was shifting, and a generation was growing up that neither knew nor cared about the days when Parkside shook with the roar of ten thousand fans singing *We've swept the seas before, boys, and so we shall again*. Many of the older supporters gave up on the club. Some went to support Leeds, now the only senior rugby league club in the city. For the remaining Hunslet fans, it felt like the club was in exile. This feeling was compounded when the owners of the Greyhound Stadium decided to demolish it. With rumours of all kinds of underhand dealings, the New Hunslet club changed ownership and had to move to share the Mount Pleasant home of Batley RLFC in 1980. The club, now known as Hunslet RLFC again, remained at Batley for three seasons before coming back to Leeds and finding a temporary home at Elland Road, which by then had been bought by the Council.

If sharing with Leeds United, the soccer upstarts, was bad enough, Hunslet faced further humiliation. Leeds United wanted the Leeds City Council to kick out the rugby league club. Hunslet wanted to return to Hunslet, and wanted the council to build them a new stadium. When the council finally agreed to do this, Leeds United insisted that Hunslet leave Elland Road as soon as possible. So, for a brief spell, while the new stadium was ready, Hunslet were forced to play out of Bramley's Maclaren Field. But eventually the new stadium – South Leeds Stadium, within sight of the ground where Parkside had once stood – was ready, and in November 1995, Hunslet, now rebranded the Hunslet Hawks, returned home. In their first season at South Leeds, Hunslet just missed out on promotion. In 1997, they appeared at Wembley in the Challenge Plate Final, losing to Hull KR. Then in 1999 they won the Northern Ford Premiership Grand Final, which was held, ironically, at Headingley Stadium. As winners of the Premiership, Hunslet should have been in line for promotion to the Super League. But intense lobbying for and against Hunslet within the game led to the Rugby Football League deciding that South Leeds Stadium was not fit for the Super League. Hunslet argued that the Council had agreed to expand the stadium if Hunslet ever reached the Super League but the RFL ignored Hunslet. The RFL's seemingly unjust decision led many Hunslet supporters to claim that the decision was influenced by the close relationship between people at the RFL and at Headingley Stadium. Many supporters stopped watching rugby league altogether, and as Hunslet's players also left the club, knowing they were denied a chance to play in the Super League, so the attendances dropped in 2000s. For the next few years Hunslet continued to struggle, and in 2007 rumours circulated that the club was about to be bought out by Leeds Rhinos or by Leeds Metropolitan University.[10] This didn't happen, but the Rhinos did send in an administrator to help the club re-organise its finances and management processes.

The rugby played by the Northern Union, what became the game now known as rugby league, permeated the lives of the working-class families of Headingley, Kirkstall, Burley, Bramley, Armley, Wortley, Holbeck, Beeston and Hunslet. The other version of rugby, played by the amateur gentlemen of the Rugby Football Union, survived in Yorkshire in pockets of middle-class suburbia, but the formation of the Northern Union in 1895 had left the Twickenham-governed weak where the northern game was strong.

Rugby League in Leeds as Globalised Super League

Leeds Rhinos were crowned the club champions of world rugby league in February 2008, following their defeat of Australian National Rugby League champions Melbourne Storm.[11] Sponsored and named after Leeds Metropolitan University's Carnegie Faculty, the World Club Championship win followed the Rhinos' Super League title in 2007. They were promoted across the city and the wider city region as one of the strongest clubs of Super League. Since the formation of the Super League and the switch to playing rugby league in summer (Denham 2004), the wealthy club of Leeds RFLC – the 'Loiners' – had been branded into Leeds Rhinos, marketing themselves as a leisure destination for middle-class families looking to fill their summer Sunday afternoons with some Americanised entertainment delivered by highly-trained, highly-paid professionals. As the Rhinos, Leeds RLFC bought into the idea of Super League as entertainment, as glamour, as the

---

10   A rumour, perhaps, bolstered by the university's buy-out of Leeds Tykes Rugby Union Football Club, which was then re-branded Leeds Carnegie.

11   Leeds Metropolitan University put out a press release celebrating this victory, available on-line at http://www.leedsmet.ac.uk/the_news/mar08/cwcc.html (accessed 15 May 2008).

northern hemisphere's equivalent of the National Rugby League in Australia. They bought the Leeds Rugby Union Football Club, re-branding it as the Tykes and themselves as Leeds Rugby – both codes in one all-year round package. With sponsorship deals and season tickets all rising year on year, the Rhinos invested in talented young players from across the city and Yorkshire, alongside the best players and coaches from Australia and New Zealand. Although the debate over which rugby league club has been the most successful in the Super League era is one repeated *ad nauseam* in pubs across the north of England, there is no doubt that Leeds Rhinos – Leeds RLFC – are on most fans' shortlists.

The change to Super League in the 1990s and increasing dominance of Leeds Rhinos in the city is regretted by nearly all the fans of Bramley and Hunslet. Arnold, a Bramley supporter since the 1960s, suggested 'Super League had given the sport publicity, but in reality is has destroyed the game ... those outside [like Bramley] have no chance now of breaking into the select group at the top'. Christine, a Hunslet fan, was critical of the Rhinos and suggested they 'have forgotten where they came from ... they were once the same as us', indicating that the Rhinos had betrayed their working-class roots and betrayed the working-class community of rugby league in Leeds. Another Hunslet fan commented that 'Leeds and others have just cast us away', expressing the feeling of a community divided and a working-class supporter base in South Leeds angry and upset at the way the game had changed.

For Charles, someone who first watched Bramley play Leigh in 1937 at the old Barley Mow ground, some of the changes to rugby league were acceptable. He was quite happy about the switch to summer and excited about attempts to expand rugby league into places like London, but he was dismissive of the commercialisation of the game. On Super League, he admitted 'they play some very good football', but went on to say that he could not

> understand why they have these names, it seems overdone ...Wakefield Trinity already had a good name that everybody knew. So why are they the Wildcats? At one time every club had a chance, now it's just the same teams.

Another Bramley supporter, Claire, echoed Charles' comments: 'I don't like all that razzmatazz; I think it's crap and too American'. She argued that Bramley's fans went to matches to watch Bramley play, to 'meet mates and have a drink and a chat'. The social world of the game was crucial to her and her view of rugby league: nobody at Bramley would watch cheerleaders or singers or fireworks because 'they would be too busy talking rugby and having a drink, or in my case selling raffle tickets'. Bramley's social world is working-class, northern, things denied by the Americanisation of Super League (Denham 2004).

Craig, a key volunteer in the Hunslet Independent Supporters Association which was set up to make a direct connection between fans and the club, also preferred summer rugby. He described a match against Workington in January as 'just freezing, far too cold', and pointed out that so many games were 'mud-baths' that spoiled the spectacle of the game. In one important sense, Craig was aware that rugby league was entertainment; its old environment and condition might not have been favourable to this function. But despite acceptance of change and modernisation, Craig was sure of the negative impact of Super League: 'Money came into the game big style but it was not shared equally... the big clubs want more and more and the rich will continue to get richer'. Craig's distrust of the big clubs and Super League extended to the Rhinos: even though he was happy for Leeds to play Hunslet in the Lazenby Cup, he 'would not go out of my way to watch them'. Another Hunslet fan was worried that the strength and dominance of Leeds in the city meant 'they might be attracting former Hunslet fans, or at least fans who in the old days might have normally gone to watch Hunslet'.

The anti-Super League, anti-Leeds sentiment was also echoed by Danny, one of the directors of the reformed Bramley Buffaloes. He believed that Super league was 'a different game [to rugby league]... players are so big, so pumped up, I think it will not be long before someone gets killed. It's not our game anymore'. Again, Danny made clear the difference between his game, rugby league, the game of the working-class semi-professional, and the Super League world of elite automata, 'pumped up' by some artificial means.[12] His hatred of Super League, and Leeds RLFC in particular, stemmed from his belief that the Leeds club had financially betrayed Bramley, that individuals at Leeds Rugby still owed Bramley hundreds of thousands of pounds. Rugby league at the home of the Rhinos was 'a totally different game [to the rugby league played at Bramley]'. Danny's anger at Leeds RLFC went further: he supported 'any team that is playing them', and suggested that the success of the Rhinos 'does nothing for us [Bramley]...if I won the Lottery I will buy Headingley and build a Netto on it!'

The only dissenting voice was Kryten, a Hunslet fan who had started watching rugby league elsewhere in West Yorkshire but who had been an active Hunslet fan since 'about eight years ago'. Kryten took a keen interest in rugby league in general and wrote about it in a number of periodicals. He supported international expansion and Super League, and on Leeds RLFC said:

> I love what they do at Headingley ... at Hunslet we have got to see Leeds Rhinos as not relevant to us, something separate ... I go to Super League games or watch it on TV when Hunslet are not playing. I love the atmosphere, the level of skill, the drama.

---

12   A reference to performance-enhancing drugs. See Waddington (2000) for the link between professionalism, commercialisation and performance enhancers.

Rugby League in Leeds: Supporting Hunslet and Bramley

Local and family relationships remain important in the lives and histories of Hunslet and Bramley fans. My own family's involvement at Bramley Buffaloes, and our previous active support of Hunslet, is an example of this intimacy (Spracklen 1996b). Robert from Bramley, one of the fans interviewed for this research, is another. His whole life has revolved around the club and the Bramley area of West Leeds. As he explained:

> I was born in Bramley, at Mount Cross [Hospital]. My grandfather was Chairman of Bramley [for many years]. I went to Broad Lane [school in Bramley], where I played rugby league. I worked for the family business [trade]. I went to my first game when I was about three or four, when my granddad took me to the Barley Mow … [I got involved because] it was in my family, with my grandfather steeped in Bramley rugby league and my dad taking me. My dad was a fan because of his dad … we were a Bramley RL family, with my brothers also keen fans.

Every male fan interviewed told a similar story of early introduction to rugby league through male members of the immediate family (mainly fathers) or, in one instance, through a close friend who attended matches with his father. The two women fans also mentioned were introduced to rugby league through men: for Christine, it was as a child because her older brother played for Hunslet; for Claire, it was through her husband. Both women shared in the working-class masculinity of the game, finding acceptance on the terraces through their involvement in various volunteering roles. But Claire recognised the male-dominated nature of rugby league's social world, its imaginary community (Spracklen 1995, 1996a, 1996b, 2001, 2003, 2005, 2007), was potentially something that hindered its development: 'the game is still male oriented generally… it is getting better and women are made more welcome these days… but that is how it has happened'. For the men, rugby league was a natural inheritance from fathers, brothers and best mates. For women, rugby league remained something to which they had to be introduced, something in which they had to prove their involvement through active engagement.

The supporters still involved and following Hunslet and Bramley were all, with the exception of Kryten, in working-class occupations: some traditional jobs, some from the new working-class of the service sector, and the rest not working or self employed in a trade. These occupations fitted working-class patterns of everyday life and their upbringing, though some had been educated in the grammar school system. All were involved in one way or another with rugby league itself beyond spectating and saw the game as a means of celebrating their working-classness. Naylor is a good example of this. A Hunslet fan since the age of eight, when he went to Parkside with his granddad, he played the game for Brassmoulders and the Station – two amateur teams based in working-class pubs, still surviving in the industrial wasteland of Hunslet, – while working as a joiner. He then went on to coach and volunteer at Hunslet Boys Club, a famous junior club responsible for its production-line of professional rugby league players. His most memorable moment in Hunslet's history was when he saw his step-son play for Hunslet at Wembley in the Challenge Plate Final in 1997. For Naylor, Hunslet, rugby league, working-class culture, working-class history and his own social world are inextricably intertwined.

In Spracklen (1996b), Sudthorpe, a northern locality steeped in rugby league, is described. Once a thriving, working-class area where people did live in back-to-back housing tucked between the pub and the factory and the rugby ground. By 1996 this locality had changed completely. The local rugby league club had been moved from the area, the terraces were gone, the factories were closed. New people lived in new houses, built on the wasteland of Sudthorpe. The white men who watched Sudthorpe dreamed of the glory days of the club, associated themselves with its past, its geography (Fawbert 2005). They became northern men by being part of rugby league. For the people who still supported the local rugby league club, rugby league and its northerness were natural – 'normal', unchallenged, essential – part of the fabric of everyday life for all the people in the past in their imagined community (Anderson 1983). Its whiteness remained invisible, unchallenged (Spracklen 2001). And just as it was in the past, so they imagined it in the 1990s. Today was the same as yesterday, the tight-knit community and family links remained the same as they have ever been; what the men did is what they had always done – the men went down the pub, watched the game, and came home to find their tea on the table. 'The Split' of 1895 was transformed as a heroic saga of working-class resistance to evil southern middle-class hegemony and control.

For fans of Hunslet and Bramley this same yearning for an imagined working-class utopia was also expressed through allegiance to rugby league's imaginary community. For example, even though the locality of Bramley has changed dramatically, and Bramley RLFC no longer play in the 'village', Robert still feels in touch with its history, its working-classness, its northerness, when he describes his own upbringing. All fans (both Hunslet and Bramley) told similar tales of belonging, of being grounded in a story of working-class, northern pride, animated by the idea that what they were doing was replicating the leisure lives of their grandfathers. Their understanding of rugby league's history was based on the orthodoxy of its own amateur historians – resistance, non-conformity, the Split, anti-RU, coloured by the memory of key moments in respective clubs' histories: Hunslet sweeping the seas before them in 1908, and so they shall again; or Bramley winning the 1973 Floodlit Trophy). As Arnold remembers:

> When we reached the Final I had to go even though it was played in the afternoon. I was working at Greenwood and Batley then, and I rang in sick. But I was spotted on the television by the people at work and got a telling off [for it] ... but it was worth it. Fantastic to win silverware. It was even better after

the match because the Directors put a spread on and invited everybody back …
for a celebration, with sandwiches and the like … we were all able to relive the
match [on television], players, Directors, supporters together in the clubhouse,
with a drink or few.

The dreams of Parkside and Barley Mow, of the working-class communities united
by a love of rugby league, ale and 'having a laugh', are turned bitter by the recent
histories of both Hunslet and Bramley. For all the Hunslet fans, including optimistic
Kryten, Parkside's glory years are a high-point against which everything else has
been measured. The closure of the original Hunslet club was seen as the fault of
directors more interested in making money out of the land on which Parkside
had been built. The exile to Batley was also blamed on individuals who, it was
claimed, sold their interest in New Hunslet to new investors who were again more
concerned with making money from the old Greyhound Stadium. Despite such
accusations of owners and shareholders caring more for money than rugby league,
most Hunslet fans were grateful for the work of anyone who had put money into
the club. Nearly all the club's sponsors and directors over the last twenty years had
working-class roots in South Leeds, or had made some money in working-class
businesses.

The refusal of the Rugby Football League to allow the Hawks into the Super
League was mentioned by most of the fans as a betrayal of the democracy of sport,
and a betrayal of the traditional working-class club: 'it all started to go wrong
for Hunslet in 1999 and we've gone backwards since then'. A couple of the fans
went further, and blamed individual directors and executives at Leeds Rhinos for
putting pressure on the Rugby Football League to stop Hunslet entering the Super
League. These same vested interests at Leeds Rugby were accused by most of the
Bramley supporters of doing their utmost to destroy the Bramley club, by buying
into the club then running it down, then lobbying behind the scenes (allegedly) to
try to stop Bramley from reforming. The energy of the supporter-led co-operative
behind the new Bramley Buffaloes club was harnessed by a desire to stop Leeds
Rhinos from erasing Bramley from rugby league altogether. The shame of having
to play out of a rugby union ground, followed by the anger of being taken over by
Leeds and playing out of Headingley, motivated Bramley's supporters to reform
the club. Its very existence proved to Bramley fans that theirs was the authentic,
working-class rugby league community, compared to the capitalists up Kirkstall
Hill. But they recognised that Leeds Rhinos and Super League remained a threat
to their continued existence. One Bramley fan was concerned, for example, that
'my son will end up a Leeds fan, cos that's all you see on the telly… we can't
compete against their marketing'. For all these respondents, the solidarity of the
game, its working-class, under-dog history, was totally undermined by the bullish
capitalism of Super League.

Rugby League: Postmodern, Globalised Spectacle?

The pessimism and nostalgic yearning of Hunslet and Bramley supporters contrast with the excitement and optimism initially associated with the commodified spectacle of Rupert Murdoch's new world order of rugby league. When News Corporation first offered millions of pounds of funding in exchange for exclusive rights to televise rugby league in 'Europe' (England) and Australia, the game of rugby league was split between loyalists (traditionalists) and those who took the cash or bought the global (expansionist) vision (Coleman 1996; Kelner 1996). In the north of England, some rugby league fans opted to defend the game against any change or expansion (Kelner 1996, Spracklen 1996b). But ultimately, in England and in Australia, the game's clubs, sponsors and administrators all accepted and welcomed the involvement of Murdoch's global media empire (Denham 2004, Collins 2006). The growth of television coverage, via News Corporation, of the European (English) Super League and the (Australian) National Rugby League has led to the game expanding into new regions and territories around the world (Spracklen 2007). In many of the developing areas, the involvement of students, attracted by television coverage, has been a catalyst to drive expansion (Collins 2006). All the respondents from the international rugby league on-line forum, by their very involvement in the game, were believers in the globalisation of rugby league: they all saw the benefits of television exposure and commercial sponsorship, and all welcomed the professionalisation of the game (Spracklen and Fletcher 2008). However, despite their enthusiasm for expansion and globalisation, they were reticent about abandoning rugby league's working-class history. Most of the respondents had some connection to the game's working-class heartlands of Brisbane, Sydney or the north of England: either they themselves were born there, or they had family there, or they had lived there for some period. As such, they saw in their expansionist work the project of a spreading the idea of rugby league as a working-class masculine game: even when they admitted that they themselves were not working class or they explained that rugby league in their particular developing country was played by middle-class students.

Clearly, rugby league is globalising, and in doing so demonstrates all the material, demographic, technological, social and cultural flows of globalisation (Appadurai 1995, Holton 2008); as well as the dominant flow of westernisation (commmodification, professionalisation, Americanisation via Australia) identified by Hall (1993), Giddens (1990) and Bauman (2000). Leeds Rhinos exemplify this globalising rugby league phenomenon, and Headingley Stadium is an outpost of the conspicuous commodification and consumption of professional sport (Horne 2006). Watching the World Club Challenge between the Rhinos and the Storm on television, with the sound of the commentators muted, one could be forgiven for mistaking the match for any other global team sport spectacle: the prominence of sponsor logos, the huge crowd in the dazzling lights of the stadium, the multicultural nature of the hyper masculine professional athletes, the designs of the jerseys and the tricks of the TV studio, all part of the Americanisation of global sport (Denham, 2004). Turn the commentary up, however, and despite the Americanised style of delivery, the rough tones and flattened vowels remind the listener of rugby league's connection to the imaginary, and the imagined, working-class world of the north of England.

There is, then, one contradiction. Rugby league is, like its union counterpart, a commodified product, its elite competitions part of the global calendar of passive consumption (Horne 2006), its international profile fuelling participative and commercial expansion into new markets. In this expansion and development, the postmodern nature of commodified sport becomes apparent (Maguire 2005). News Corporation and other multinational sponsors create supranational leagues, rugby league clubs change, lose their local identity and become businesses in the same way as elite football clubs have changed (Fawbert 2005). As Denham (2004) has argued, rugby league's embrace of Americanisation and commodification, a postmodern turn itself, is evidence for the dissolution of identity and liquidity of structure (Bauman 2000) associated with postmodernity. The existence of Leeds Rhinos and the World Club Challenge provides evidence for some postmodern shift away from high modernity (Giddens 1990), from the traditional, fixed working-class communities and identities typified by the Hunslet of Richard Hoggart (1957). Leeds itself is a microcosm of the move towards postmodernity and the decline of the traditional, industrial base of the working-class economy. The globalisation of rugby league demonstrates some embrace of the rationality of instrumental capitalism at the end of modernity (Habermas 1987), perhaps the beginning of postmodernity (Harvey 1989; McGuigan 2006). But in Hunslet and Bramley, individuals still choose to identify with working-class communities of modernity, expressing communicative action (Habermas 1987) in resisting the conformity of Super League. Through the imaginary community of rugby league, these fans reject the idea that rugby league is just another sport: for them, it is a way in which they identify totally with their white, northern, working-classness. The game gives them a sense of belonging, a sense of place, a fixed point around which they have constructed their identity and the identity of everyone else who belongs to rugby league in Hunslet and Bramley (and beyond, into the imaginary world where the north ends at York and Newcastle does not exist). In their active support of Hunslet and Bramley, these fans choose to be identified through structures of class, gender and ethnicity. They choose to distance themselves from the razzmatazz of the Rhinos, the corporate show of Super League, though they mourn the disconnection between their rugby league and the rugby league of the elite. By belonging to the imaginary, imagined community of rugby league, by claiming continuity with the past and by re-enacting that past in the present through active engagement with Hunslet and Bramley, these fans are challenging the impact of postmodernity on identity formation. Whatever might happen at Headingley Stadium, no matter how many Ronnie the Rhino screensavers are downloaded from the Leeds Rhinos web-site, the active supporters of Hunslet and Bramley will remain comfortable and secure in their (re)constructed, imagined, working-class identity.

References:

Appadurai, A. (1995). *Modernity at Large: Cultural Dimensions of Globalization.* Minneapolis: University of Minnesota Press.

Anderson, B. (1983). *Imagined Communities.* London: Verso.

Bauman, Z. (2000). *Liquid Modernity.* Cambridge: Polity.

Collins, T. (1999). *Rugby's Great Split.* London: Frank Cass.

Collins, T. (2006). *Rugby League in Twentieth Century Britain.* London: Routledge.

Colman, M. (1996). *Super League: the Inside Story.* Sydney: Pan Macmillan.

Denham, D. (2004). Global and Local Influences on English Rugby League. *Sociology of Sport Journal,* 21, 206-19.

Fawbert, J. (2005). Football Fandom, West Ham United and the 'Cockney Diaspora': from Working-class Community to Youth Post-tribe? in *Sport, Active Leisure and Youth Cultures,* edited by P. Bramham and J. Caudwell. Eastbourne: Leisure Studies Association, 171-94.

Fletcher, R. and Howes, D. (1995). *Rothman's Rugby League Yearbook 1995-96.* London: Headline.

Giddens, A. (1990). *The Consequences of Modernity.* Cambridge: Polity.

Habermas, J. (1987). *The Philosophical Discourse of Modernity.* Cambridge: Polity.

Hall, S. (1993). Culture, Community, Nation. *Cultural Studies,* 7(3), 349-63.

Harvey, D. (1989). *The Condition of Postmodernity.* Oxford: Blackwell.

Hoggart, R. (1957). *The Uses of Literacy.* London: Chatto and Windus.

Holton, R. (2008). *Global Networks.* Basingstoke: Palgrave.

Horne, J. (2006). *Sport in Consumer Culture.* Basingstoke: Palgrave.

Hughson, J. and Free, M. 2006. Paul Willis, Cultural Commodities, and Collective Sport Fandom. *Sociology of Sport,* 23, 72-85.

Kelner, S. (1996). *To Jerusalem and Back.* London: Macmillan.

Maguire, J. (2005). *Power and Global Sport.* London: Routledge.

McGuigan, J. (2006). *Modernity and Postmodern Culture.* Maidenhead: Open University Press.

Spracklen, K. (1995). Playing the Ball, or the Uses of League: Class, Masculinity and Rugby, in *Leisure Cultures: Values, Genders, Lifestyles,* edited by G. McFee et al. Eastbourne: Leisure Studies Association, 105-20.

Spracklen, K. (1996a). When you're Putting yer Body on t'line fer Beer Tokens you've go'a Wonder why: Expressions of Masculinity and Identity in Rugby Communities, in *Sport, Leisure and Society Vol. One,* edited by G. Jarvie et al. Edinburgh: Heriot-Watt University, 131-8.

Spracklen, K. (1996b). *Playing the Ball: Constructing Community and Masculine Identity in Rugby.* Unpublished PhD thesis: Leeds Metropolitan University.

Spracklen, K. (2001). Black Pearls, Black Diamonds: Exploring Racial Identities in Rugby League, in *'Race', Sport and British Society,* edited by B. Carrington and I. McDonald. London: Routledge, 70-82.

Spracklen, K. (2003). Setting a Standard?: Measuring Progress in Tackling Racism and Promoting Social Inclusion in English Sport, in *Sport, Leisure and Social Inclusion*, edited by I. Ibbetson et al. Eastbourne: LSA, 41-57.

Spracklen, K. (2005). Re-inventing "the game": rugby league, 'race', gender and the growth of Active Sports in England, in *Sport, Active Leisure and Youth Cultures*, edited by P. Bramham and J. Caudwell. Eastbourne: LSA, 153-67.

Spracklen, K. (2007). Negotiations of Belonging: Habermasian stories of minority ethnic rugby league players in London and the South of England. *World Leisure Journal,* 49 (4), 216-26.

Spracklen, K. and Fletcher, T. (2008). They'll never Play Rugby League in Kazakhstan: Expansion, Community and Identity in a Globalised and Globalising Sport. Paper presented to the Leisure Studies Association, Liverpool, July 2008.

Wheaton, B. ed (2004). *Understanding Lifestyle Sport.* London: Routledge.

Chapter 9

"Off with their Headscarves, on with their Football Kits?": Unveiling Myths and Exploring the Identities of British-Muslim Female Footballers

Aarti Ratna

Introduction

In 2003, in Australia, a women's football game was due to be played between Melbourne Women's Soccer Club at Keilor Park (BBC News 2004). Before the match was allowed to start the referee forbad one of the female players from taking the field of play. The player in question was wearing a hijab, a headscarf worn by Muslim women. The referee asked her, "Is something wrong with your head?" (Turnbull 2007). Afifa Saad, the player in question, refused to take off her headscarf at the referee's request and as a result of this altercation the match had to be postponed. Afifa asked the official to explain to her how wearing a hijab affects a woman's ability to play football (BBC News 2004). Eventually the Victorian Soccer Federation intervened and ruled that Afifa should be allowed to play in her headscarf. In other cases, closer to home, Ansar Women's Football Club of East-Glasgow – a predominantly Muslim team – was prevented from playing matches by the Scottish Women's Football Association because some of the players chose to wear a headscarf (Turnbull 2007). The rulings of these separate cases are based on interpretations of relevant rules of the Fédération Internationale de Football Association (FIFA). In particular, Law 4 is about compulsory equipment and the minimum gear needed to play the game; it implies that only 'basic' equipment is allowed to be worn but also stipulates that players are able to use other 'non-basic' items such as headgear, protective wear and gloves (Cobb 2007, Turnbull 2007). Although some goalkeepers are allowed to wear baseball caps, the wearing of the hijab is questioned. It becomes apparent that what is defined as 'basic' and 'non-basic' is left for the ruling bodies in different countries across the world to interpret as they see fit. The law specifically fails to show any sensitivity to the 'basic' needs of dress that some followers of Islam choose to follow (Turnbull 2007).

Debates about Muslim females and the headscarf are not confined to sport. In France, for example, donning the headscarf at school has come under intense public and political scrutiny. Such controversy has exposed long-standing tensions between those who want to secularise the state and the education system and those who see the banning of the headscarf as a perversion of the human rights of Muslim women (Bowen 2006, Wilson 2006). In 2006 in the UK the then Foreign Secretary, Jack Straw, angered Muslim groups by suggesting that the veil could be a 'visible statement of separation and of difference,' and that he would ask women visiting his surgery in Blackburn to consider removing it (BBC News 2006). In the context of urban unrest in towns and cities across the North-West of England following the summer of 2001, many disaffected young Muslim men and women felt that this viewpoint did little to embrace them as citizens of the British nation. It was felt by many that they continued to be regarded as 'the enemy within' (Abbas 2005) and moreover, that Muslim women in particular were still seen as 'victims' of the Islamic way of life (Wilson 2006). Arguably, in postmodern cities of the north-west such as Blackburn, Oldham, Bradford and Leeds – home to a high proportion of Muslim people in England – the celebration of multiculturalism and difference appears as tokenistic when set against such confrontational political rhetoric and actions. Such positions are ideologically reproducing state-based racism and, at the same time, ignoring the impact of such policies on the lives, perceptions and experiences of Muslim people living in these areas. Furthermore, and importantly, what becomes apparent is that wearing a headscarf bears a symbolic weight out of proportion to the number of women actually wearing it (Bowen 2006). It is a stigmatising signifier in a racist discourse.

The British-Muslim Diaspora

Like many cities in England, Leeds is known for celebrating and embracing the rich diversity of ethnic groups living in the area (Stillwell and Phillips 2006). However, in the context of football, Leeds is noted as the most racist club in Football League despite attempts by supporter organisations to challenge and change behaviour. According to the population census in 2001, 15,064 of the overall population of Leeds are of Pakistani origin (Office for National Statistics, 2001). Although it may appear that the population is relatively small in comparison to the 67,994 Pakistanis in Bradford, they still represent a sizeable population within the region of Yorkshire and Humber (146,330). Brah (1996) argues that many South-Asian[1] women followed their fathers and husbands from the Indian sub-continent to live in parts of England. She argues that there may be a myriad of reasons to explain their migration destinations including the legacy of colonialism, the labour market conditions of the area of settlement as well as expectations related to education and employment. In parts of the North-West and North-East of England as the growing textile industry became more rationalised in the post-war period, many

---

    1   The term South-Asian is used to define groups of people from the Indian sub-continent.

factories and mills began to become competitive traders on the world market (Kalra 2000). This in complex ways can be linked to the decline of such industries in parts of the Indian sub-continent. Indeed, many cotton manufacturers were no longer able to sell their products to internal markets as it became cheaper to import from countries such as England. Hence, thousands of Pakistani, Bangladeshi and Indian families were forced to migrate away from their homes in order to earn a living in Britain. Through this shift, ironically, many of the migrants settled in rural parts of Lancashire and Yorkshire working in the mills that had initially made them unemployed. In Britain, as the textiles industries declined in the 1980s and 1990s, many of the migrant families moved again, this time to urban centres such as Leeds. Kalra (2000) documents how many families left the mills to became taxi drivers in urban areas of Lancashire and Yorkshire and often found themselves living in the poorest areas of the city.

Multicultural Cities and the Representations of British-Muslims

In recent times, Leeds' status as a harmonious multicultural city has been questioned especially following the disturbances in 2001. An investigation into the urban unrest, led by Professor Ted Cantle, reported that so-called 'parallel lives' and ethnic segregation of local communities had generated deep-rooted divisions within the city (Parekh 2006). Trevor Phillips, the chair of the Commission for Racial Equality (CRE) in England, has argued that such divisions are indeed reflective of increased ghettoisation of ethnic minority communities (Phillips 2006). He believes that different ethnic groups share spaces in towns and local cities but know very little about each other and he argues that as a country England is 'sleeping walking' its way towards the scale of segregation seen in the Unites States of America (BBC News 2005).

In the spaces of sport, it becomes apparent that citizenship and multiculturalism are narrowly defined; a popular belief is expressed in the notion that 'you are either for us or against us' (Bagguley and Hussain 2005). In the report by the Runnymede Trust (2000) – also known as the Parekh Report – about the *Future of Multi-Ethnic Britain*, it is argued that being British is racially-coded. Britishness is linked to Englishness and only those who are included within that imagined template are considered legitimate citizens of the state. In contrast, those who do not fit within the symbolic boundaries of Englishness are considered to be outsiders to the national community (Fortier 2005). In the wake of such findings, Stuart Hall (2000) argues that the 'multicultural question' continues to loom over politicians. He feels that these politicians must take stock, contend and re-define British citizenship with regard to the positions of ethnic minority groups within the nation state so they no longer continue to be alienated outsiders (see also Fortier 2005).

The Limits of Multicultural Leeds

Phillips *et al.* (2007) explore the way that Leeds (and Bradford) is imagined, experienced and also understood by British-Asians, including those of Muslim heritage. From Census data (2001) about ethnic minority groups in Leeds, it appears that since the Second World War inner-city ethnic minority concentrations in places such as Harehills, Beeston and Hyde Park have strengthened. Evidence indicates that approximately 60% of British-Muslims are living in the poorest areas of the city (Stillwell and Phillips 2007). Phillips *et al.* (2007) argue that despite recent concerns about ethnic segregation, a growing number of Bangladeshi, Pakistani, Indian and Kashmiri groups are moving to the suburbs, resulting in a more complex neighbourhood mixture of racial and ethnic groupings. For example, increasing numbers of British-Asians have moved to desirable neighbourhoods within Chapel Allerton, Alwoodley and Moortown – generally seen as aspiring places for upwardly mobile residents of the city. Interestingly white ethnic segregation in parts of East Leeds, including places such as Crossgates, Middleton and Seacroft, is not seen as problematic but 'normal' (see Bonnett 2000). However, in Phillips *et al.*'s (2007) study the authors identify that for many British-Muslims in particular, their experiences continue to be limited to such ethnic enclaves. Interestingly, many of the British-Muslims living in such neighbourhoods would argue that the supportive community and religious networks available to them in these neighbourhoods ensure their well-being in an otherwise alienating city. In Leeds, these positive aspects of British-Muslim residences are ignored in favour of a common perception based on the idea of an insular community which suffers from problems of social exclusion and in desperate need of help. Unsurprising, such urban spaces are at times seen as threatening. For example, in the 1980s many British-Muslims in Leeds (and Bradford) where seen as fundamental Islamists, especially during the aftermath of the Rushdie affair[2] (McLoughlin 2005). In more recent times, ethnic enclaves have been seen as hotbeds of terrorism as three of the four young men to bomb London in July 2005 had lived in Leeds.[3] Through such media and state depictions about the ethnic clustering of British-Muslim communities, the segregation of British-Muslims is often linked to their 'backward' religious and cultural differences rather than other possible institutional constraints on their lives and experiences. For example, racial segregation can be linked to poor access to council housing and employment opportunities and so limit ethnic minority patterns

---

2   In 1989, Ayatollah Khomeini – the spiritual leader of Iran – issued a fatwa requiring the execution of the fiction writer Salman Rushdie. The publication of Rushdie's book, *The Satanic Verses*, had sparked controversy over the perceived irreverent depiction of the prophet Muhammad. In several western countries, many Muslims people held public rallies in the streets in which copies of the book was burned (Werbner 2000).

3   They are usually referred to as the three Leeds bombers–Mohammed Kahn was from Beeston and had recently moved to Dewsbury, Hanib Hussain lived in Holbeck in Leeds and Shelzad Tanweer lived in Beeston, having been born in Bradford; Germaine Lindsay was a Jamaican, lived in Huddersfield and then moved to London.

of residency across Leeds. Interestingly, Phillips *et al*. (2007) suggest that, although some young British-Muslims are leaving the city in search of jobs and education in other parts of England, the mobility of many younger generations of British-Muslims is restricted by the continued prevalence of institutional exclusion, racist harassment and feeling out of place. Indeed, many areas of Leeds are effectively off-limits to British-Muslim young people: conversely pre-dominantly British-Muslim areas are seen as no-go areas for white residents.

British-Muslim Women and the City

Asfar (1994) argues that in the wake of the Rushdie Affair, British-Muslim women in Yorkshire shared a sense of political unity with their men folk, struggling to cope with life in a hostile city. Asfar (1994) also suggests that at this time, Muslim women in the region were perhaps more vulnerable to social exclusion as they mainly worked from home and had little access to health care and welfare facilities in the city. Ali (2000) argues that in many cities in West Yorkshire, local politicians often viewed the problems of Muslim women – relating to dress, modesty, patriarchal relations and other religious observances – as internal affairs for the community to deal with itself. Indeed, Ali (2000) argues that such a policy implies that cultural differences are respected and are not seen to be part of a wider dialogue about social, economic and political relations in society in general. In more recent times, Phillips *et al*. (2007) found that many British-Muslim women feared crossing from Asian to white spaces within the city. Hussain and Bagguley (2007) stress that girls and young women in Leeds suffer from multiple forms of deprivation which relate to education, income and employment and result in feelings of exclusion, isolation and discontentment. In light of such issues, many young British-Muslim women are turning to Islam, and donning of the hijab, in order to find a sense of belonging in an otherwise unstable place (Ali, 1992).

Hargreaves (2005), Scraton (1994) and Watson and Scraton (1998) suggest that, even though we may be moving into a postmodern age, the impact of common structures of constraint relating to gender, 'race'/ ethnicity, sexuality and class, in sport and leisure, are still ignored. Moreover, Scraton (2001) argues that the voices, needs and experiences of both black and South-Asian women are relatively ignored within both academic and policy discourses about social inclusion, multiculturalism and gender equality. The English Football Association suggests that it wants to use the 'power of football' to encourage more people – regardless of 'race', religion or background – to get into the game (FA 2009). Yet, this does not mean that Muslim female football players are positively embraced in mainstream settings of women's football or that they are recognised and valued as players of the game (Ratna 2008). Many experiences and perceptions of British-Muslim females living in Leeds resonate in many other urban areas of England and in national discourses on segregation, community cohesion and the multiculturalism. It is to this wider contextualisation that we now turn to in order to consider how

British-Muslim females negotiate and assert their rights to play football locally, nationally as well as internationally.

A Note about Researching 'Race' and Ethnicity

The analysis of 'race' has been a significantly under-researched area of both sport and leisure studies (Carrington and Bramham, 2007; Carrington and McDonald, 2001). Moreover, connections between 'race' and gender have focused on the lives of South-Asian and/or black men rather than their female counterparts (Scraton, 2001). This chapter offers an original insight into the identities and experiences of British-Muslim females in particular. Moving away from pathological and ethnocentric interpretations of their lives, cultures and sporting preferences (Raval, 1989; and Zaman, 2001), in this section I aim to address how young women accommodate, negotiate and challenge discourses and structural constraints in and through their participation in football. Moreover, I attempt to link British-Muslim females experiences to wider relations of power and the unequal distribution of resources/knowledge about football to issues related to team subcultures and individual experiences of the sport, as well as the cultural values and norms associated with the spaces of (white) women's football.

The interviews used in this section stem from my own doctoral research about British-Asian females and football (see Ratna 2008). Here I will particularly focus on the voices of six British-Muslim players who mainly play for the UK Muslim women's futsal team and/or have represented university clubs and/or other women's football teams. In alphabetical order, they are Farrah (26 years old), Hani (21 years old), Mieka (29 years old), Saima (21 years old), Shari (19 years old) and Zenab (18 years old). They are from different parts of the country including Leeds and other areas in the north-west of England. Additionally, the opinions of other Hindu and Sikh British-Asian females who play football in this country are used in order to explore their opinions about the involvement of their Muslim female counterparts.

The identities of British-Muslim females are considered to be heterogeneous, structured as they are by gender, 'race'/ethnicity, religion as well other pertinent markers of social identities which include age, generation, class and sexuality. Brah (1996) suggests that exploring the intersections of such identities, and differences between women who may identify with the British-Asian label, is central to the development of a more sophisticated understanding of their multiple experiences and subjectivities. Interestingly, as Modood (2005) notes, religious affiliation is prioritised in the British-Muslim label or category. Arguably, many young Muslim men and women may see themselves as British citizens by virtue of the fact they were born in the UK. Moreover, members of the Muslim community choose this label in order to find a place of belonging in the otherwise alienated spaces of this country (Modood 2005). Dwyer (2000) specifically utilises this conceptualisation of British-Muslim in her study to explore the role that gender plays in the construction and negotiation of this identity.

British-Muslim Females and Sport: In the Name of Allah

Lovell (1991: 166) argues that the roles of South-Asian females in Britain – including those of Muslim religious identity – are bound by traditional ideas about femininity which involve staying at home and learning how to cook and clean in preparation for becoming a wife and mother. Farrah, who represented her local school team, suggests that, in her opinion, '… in the South-Asian culture it is expected for girls to be feminine/girlie, playing sports is seen as outside the norm'. The norm she is referring to is to stay at home, get married, have children and also to meet moral and religious expectations in terms of dress, such as to appear modestly. Walseth (2006: 91) also found in her study of young South-Asian Muslim females that played sport in Norway that ideal notions of femininity entail that they should not smoke, drink alcohol or go to places where men are present. Many of the British-Asian Sikh and Hindu players included in my research actually promoted the idea that even though *they* may not be repressed by their cultures, British-Muslim women in contrast are indeed prohibited from taking part in a whole range of 'non-Muslim' cultural activities. For example, a female of Sikh-Indian heritage, who has played for a well-known women's football team and who also coaches African-Caribbean and British-Asian primary school-aged girls, suggests that:

> I have a lot of Asian friends and I do ask them. They seem to find football boring but others are restricted. Take Muslims for example. They can't show their legs so football is out of the question for them. If it is a problem they should bring it up y'know, 'Are we allowed to wear tracksuit bottoms?' I used to have this Muslim friend who I used to play [football with] at secondary school, she used to love playing football … she was like 'I'm more free when I am at school because I know I can play there, but if my family sees me play I don't think they will let me go ahead with it'. Now showing her legs was one of the reasons for that, y'know some families are more supportive than others … . (British-Asian football player of Sikh-Indian heritage)

To explain her viewpoint it is important to acknowledge that, at this time, British-Asians *as a whole* are increasingly viewed as 'the enemy within' in media depictions (Abbas 2005, Modood 2005). Although, all British-Asians are not scapegoated in the same way, it becomes apparent that Sikh and Hindu British-Asians are more likely to be viewed as 'good' citizens and to have been 'assimilated' to mainstream British life fairly well, breaking through in education and employment and contributing positively to the British economy (Ahmad 2001; Ahmad and Modood). As a result of their success, it is often suggested that in postmodern cities, inequalities belong only in the past, and with the right talent and determination, ethnic minority people can aspire to and obtain positions of authority and power. In contrast to the Sikh and Hindu British-Asian communities, Muslims, mainly from Pakistan and Bangladesh, are often viewed in a negative light. It is assumed that, for a number of reasons linked to their own cultural inadequacies and self-imposed isolation, they have not made such a mark. Such a viewpoint renders invisible the number of Muslim men and women who worked in the mills of northern England and contributed to the British economy in the 1960s (Kalra 2005). Some of these men, with little opportunities for jobs in other blue and/or white collar sectors, are highly visible in many northern towns and cities as taxi drivers (Kalra 2005). Additionally, those Muslims who do not support the invasion of Iraq and America's 'War on Terror'[4] are considered to be 'bad' citizens (Abbas 2005 and Wilson 2006: 56). Politicians are quick to remind the public that the 7/7 bombers were indeed 'home-grown', three of the men in fact grew up, as I observed earlier in the Leeds district of Beeston. With such popular opinion in mind, it is easier to understand why some Hindu and Sikh females may be keen to stress their similarities to the white residents of England and in comparison their differences from the 'backward', materially disadvantaged British-Muslim community. Furthermore, as indicated in the above testimony, Muslim female players may be 'blamed' for not taking responsibility for their own participation needs and requirements. Through this strategic way of thinking, some Hindu and Sikh British-Asian female players re-work discourses about femininity in order to position themselves as relative insiders to women's football, compared to Muslim British-Asian players.

However, many of the Muslim players themselves claim the opposite. They suggest that, if opportunities were available, practising Muslims would happily play sport at school and for women's football clubs. Some of the players included in this study were selected to play in the 2001 Muslim Women's World Championships and/or were training to play in the 2005 tournament. Mieka – the captain of the 2001 team (and assistant coach of the 2005 squad) – explains that people she worked with were surprised when they found out she played football. She believes that:

> It is not the first impression people have when they see you, a Muslim woman in Muslim dress … that was another great thing about the Games, it broke down stereotypes, it showed that Muslim women are not chained … we can participate in all kinds of activities in everyday life.

Many of the Muslim females that play for the UK women's futsal team discuss how modern interpretations of the Koran advocate female equality with males. Female Muslim players use their own religious discourse as a form of cultural power to justify and legitimise their football participation. Many of the British-

---

4   At the beginning of 2009, David Miliband is seeking to distance himself from this discourse as it gives an impression of Muslim people as a unified global enemy. This phrase also suggest that the correct response to this is a military one rather using the law to foster co-operation between the state, different organizations and ordinary people (Miliband, 2009).

Muslim players interviewed for this study echoed the words of Saima – one the players who was due to represent the UK team in 2005 – who states:

> ... our book, the Koran, is full of sayings and things to do with the prophet, and that's what we go by. The Koran suggests that the prophet used to race with one of his wives quite regularly and he used to get upset if he didn't win. Y'know it was a proper competition, he used to encourage her to be fit in that way. That's just one example, I'm sure there are other examples of the same thing. But there is nothing in the Koran [to suggest] that women aren't allowed to do sports and they aren't allowed to enjoy themselves.

Dagkas and Benn (2006) believe that there have been some important modernizing changes in recent years and women are interpreting the Koran in a new progressive light. Some Muslim women choose to wear the hijab and others do not, but this does not necessarily mean that those wearing the hijab are more devout than their team-mates who choose not to. Furthermore, many Muslim females choose to adopt the hijab even if their mothers did not. They maintain that wearing the hijab is both a personal and a political choice and that it helps them develop a sense of community and it is not a form of patriarchal oppression.

Off with the Hijab or not?

Wilson (2006: 24) notes that during these increasingly unstable times, following 9/11 and 7/7, many Muslim females are choosing to wear the hijab as a political choice to express their opposition to anti-Muslim sentiments in Britain and other parts of the world. Shari points out that there are a great many differences between Muslim women, but, 'in the west, we are all perceived as '*Muslim*' women and therefore we are the same'. The differences and diversities between Muslim women and their unique identities and experiences are often ignored. Shari suggests that such differences – in terms of ethnicity, religious affiliations, belonging to various sects, geographical variations in residence, and age, for example – affect the sporting experiences of Muslim women in a myriad of ways that are not appreciated by policy-makers and/or sport providers in the UK. Shari is claiming that homogenising British-Muslim females fails to appreciate how identities are multiple and shifting, influenced by the rapidly changing cultural norms and values of shifting patterns and experiences of socialisation. Adopting a feminist discourse, British-Muslim female players affirm and assert their heterogeneity in relation to each other as well as to other young women in England and the rest of the world. Similarly, it would be wrong to assume that Islam is the only influencing cultural force in their lives. For example, dominant western ideologies about femininity are incorporated into ideas about why Muslim women (and men) should be physically active. Mieka states:

> Well if a guy comes home and his wife is not very fit and healthy he ain't gonna
> be happy. You know what guys are like these days, the way they want the perfect
> size and model-size female. I mean it is a kind of give and take, as girls want
> guys to be fit and healthy … so going to the gym is a must!

Likewise, Muslim females do not necessarily choose to dress traditionally in
the hijab. Many Muslim women are increasingly fashion-conscious and they are
able to mix designer cuts and styles with certain Islamic forms of dress and/or
buying specially-cut western clothes that preserve their modesty, for example
long-sleeved Gucci or Armani tops. Mieka adds that it is wrong to think Muslim
females 'don't care about the way that they look'. She points out that she likes
wearing sporty clothes and Islamic dresses as well as western clothes such as jeans
and t-shirts: for example, 'Look what I am wearing today [headscarf with Calvin
Klein logo across the top and shalwar kameez[5]], I am not wearing sporty clothing
now'. Khan (1993: 68 cited in Hollows, 2000: 157) suggests that 'British' and
'Asian' are not exclusive cultural categories that form separate identities but rather
are fused to form a hybrid style. Modood (1997: 338) suggests that 'the ways in
which minorities conceive of themselves and the cultural synthesis that is taking
place are various, changing and generating new and mixed forms of ethnicity'.
The consumption and adoption of both Islamic and western forms of dress are
embraced by the British-Asian women as one means of empowerment as well as
facilitating their participation in the game of football.

Representing 'England': Belonging and Inclusion

Some of the British-Muslim players who reflected upon the idea of playing for
India and/or England echoed the idea that they do not really belong to either nation.
For example, Saima, who played for the UK futsal team at the Muslim Women's
World Games in Iran 2005, describes this lack of identification:

> I think if I played for this country [England] they would see me as an outsider
> as I'm Asian, and, if I play for my parents' country which is Pakistan, they
> would also see me as an outsider because I'm not really Pakistani, I'm British.
> I probably can't relate to their culture or their experiences or anything like that
> … I mean, I'm Pakistani because my parents were born there but I wasn't born
> there and I haven't lived there … I'd probably be more *cultural* if I was brought
> up there [emphasis added].

This player views herself as different from people born in Pakistan, in part because
she believes that there are traditional 'cultural' dogmas in Pakistan that restrict the
freedoms of other Muslim girls and women. In comparison, she feels that she has

---

5   The shalwar kameez is the Muslim dress of a long tunic with trousers.

more freedom to play football as she has been able to adopt a more sophisticated and modern interpretation of Islam. In essence, Saima distinguishes between British-Asian Muslims whom she sees as modern and Pakistani-Asian Muslims whom she considers to be backward.

The sense of separateness that comes from not belonging in England, and likewise not belonging to your 'home' away from 'home' in the Indian sub-continent, may make some of these players feel like they do not belong anywhere. Some of the Muslim females that I interviewed resolve this tension by finding an alternative space, where they do belong and where they can reconcile their British and Asian identities without fear or constraint. Bhabha's (1994) notion of 'third space' is important as he suggests that the creation of an alternative space does not simply create an environment where people can celebrate their own hybridity but is also a space that is politicised. Moreover, the identities and experiences of the British-Muslim female footballers, reveal their agency through the actions they take to facilitate their own participation in response to constraints in the mainstream organisation of football in England and also to their marginal positions in this country. Mieka, the captain of the Muslim women's futsal team in 2001, explains how playing for the UK enabled her specifically to link her British and Muslim identities in a way that would not have been possible if she had played for the England Women's Football team. Zenab further articulates the following point:-

> I think it was the first time … as I have always considered myself as a British-Muslim, the two kind of went together…I'm British and I'm Muslim and it was the first time I was able to represent both of those in the same place …. y'know, I'm kind of joking now, that my fame was to tell everyone that I played football for England kind of thing y'know? I mean without the Muslim World Games for Women, I don't think I would have had that opportunity to say that.

Generally, the girls are full of praise for the organisers of the Games in Iran and for the UK Muslim News who were responsible for bringing together their squad and mostly they felt proud to identify with the other Muslim women they met from across the world. The former captain of the team describes the grandeur of the Games and her sense of belonging:

> I remember the opening ceremony was just like the Olympic Games, but on a smaller scale, with fireworks. There was a display, all the teams walking out in this huge stadium, and everyone was invited and it was absolutely fantastic, the feeling was so powerful, just walking with all these Muslim women from different parts of the world and we're all united by the fact we're Muslims, and there for the Games, so that was fantastic. (Mieka)

Mieka echoes the sentiments of the other Muslim British-Asian players I interviewed for this study, she believes in 'Umma' – a shared sense of community with other

Muslim women from across the world. In the case of this study, despite their pride in reconciling their British and Muslim identities, a closer look reveals that their feelings of belonging were actually characterised by mixed emotions.

The Muslim women's team in this country has not been able to secure funds from the FA, the governing body of the game, although the British Embassy has previously sponsored them. Yet despite securing some funds, in the words of Mieka who was the captain of the team at that time, the government perhaps had ulterior motives rather than a genuine desire to help the UK futsal team:

> I mean it was 2001 just after the Afghanistan bombing, so I think the Foreign Office got involved as they wanted to show that we can build bridges between Britain, the Middle East and Muslims all over England. I think they did this to show that the war wasn't about an attack on Islam or Muslims per se, it was about what was happening there.

Many of the players – Mieka, Zenab, Saima and Shari – felt it was a political strategy deployed as a token gesture to disprove claims that British politicians, and the public at large, were anti-Muslim. The players considered that this did little to embrace Muslim communities living in England as the Games were not well publicized and they still felt like 'outsiders within'. With limited resources and facilities they have organised their own football sessions in public places such as parks, school halls and various outdoor venues in order to train as much as possible prior to the tournament. The public venues meant that some of the Muslim females consciously chose to wear headscarves whereas some of the other players do not wear a headscarf at any time on or off the football pitch.

'West is the Best'?[6]

The increasing number of divisions between British-Asian females in terms of their religious affiliations, not forgetting other pertinent markers of difference, means that political activism regarding British-Asian football participation is fragmented rather than concerted. For example, the Muslim Women's Sports Foundation have formed their own political organisation to raise awareness about their particular concerns, wishes and beliefs and this remains separate from the larger political work of the Asian Football Forum working group, who campaign on behalf of Muslim, Hindu and Sikh (male) players. Their agendas and motives are similar in some cases and different in others. It may be argued that the MWSF, as a separate political body, gives them a platform to voice their unique experiences, needs and wishes.

---

6   'The West really is the Best' was the title of Polly Toynbee's article in The Observer, 02.03.2000. She developed these arguments later in her chapter, 'Who's afraid of Global Culture?' in Giddens and Hutton's (2000) book, *On the Edge: Living With Global Capitalism.*

These futsal players set up the Muslim Women's Sport Foundation in England as a vehicle to dispel specific myths about Islam and other cultural stereotypes (Sporting Equals, 2006). One reason for this collective action is to rectify the belief that 'the west is the best' (Shari) in terms of sports facilities and services. As a result of playing in Iran at the Muslim Women's World Games, many of the players recognize that more could be done in England to support the sports participation of Muslim women. In Iran, women's national sports teams have 'top-of-the-range' facilities, equipment, medical support and expertise to guide success on the international stage. In an interview with BBC News in 2001, Malika Chandoo, then captain of the Muslim UK squad, thought that 'West was the best' in terms of sports facilities and opportunities for women to play sport (BBC News 2002). However, after seeing the high-level organization, facilities and support given to Muslim women in Iran, she is now more sceptical. Hani, another player from the 2001 UK futsal squad, and referring to a group of Islamic girls and women that she coaches at the weekend, suggested that she found it really hard to find somewhere affordable for them to train. Furthermore, the cheaper facilities were heavily booked-up all year round. Hence, they train at a local park despite the unpredictable weather in England. The same problems of resources are experienced by the organizers of the Muslim women's futsal team. Hani goes on to explain that:

> One thing stopping Muslim women (participating in sport) is that they just haven't got the facilities to play at. One of the things that stop us from playing as regularly as we would want to is funding. We don't have the equivalent facilities or equivalent funding (as the England women's football team), and when we finally do get it, the cost of hiring is still beyond our price range.

The futsal team train together once-a-month and then every week closer to the time of the tournament. Training takes place in the south of England, which is easiest for most of participants to get to. However, they have to pay their own travel costs and ultimately their own airfares to Iran. Even though they are representing England they have received no funds from the FA in either 2001 or 2005. For many of these British-Muslim footballers this is a real financial burden. Additionally, they rely on the voluntary coaching support of a woman. The editorial team at the 'Muslim News' organize and support the futsal team as well as the other UK squads – basketball and badminton – and pay for most of their other expenses like kit and hiring of equipment and facilities. Listening to the British-Muslims' stories, it becomes clear that despite their desires to play football at local and international levels, their participation is hindered by a variety of non-cultural and non-religious factors.

Community Sports

Zenab, who considers herself to be the only Muslim female that plays football in the part of England where she lives, suggests that 'my level of practising Islam is probably not as high as some other people's'. She explains that her low level of football training is due to very few opportunities for Muslim women to play segregated football where she lives in Leeds. She uses the opportunities for playing football that are available even though she knows that to do so is a cultural taboo in many Muslim communities. So, she plays with a group of 'lads' and also takes part in the training sessions for the Muslim Women's World Games. Zenab did not view mixing with males in order to develop her football skills as problematic, even though many British-Asian and white observers – for different reasons – would consider mixed football to be unsuitable for British-Muslim girls.

In the last few years a number of young British-Asian teams have begun to form in and around Leeds including the self-named 'Leeds Wildcats'. In 2007, members of the Leeds Wildcats were featured in a special report during the BBC's broadcasting of the Women's FA Cup Final. These girls started the team a number of years ago and strongly articulate the view that, despite the perception the public may have of them, they can and they do play football. The confidence of the team partly stems from their collective agency to tackle disadvantages in the area where they live, mainly Beeston and the surrounding neighbourhoods of south Leeds. Since the formation of the club, the females have played in competitions across the country including the FA and BBC's joint initiative 'Your Game' (BBC News 2008). This project is aimed to help young people in under-represented communities become involved in football, music and the media. In 2008, the Leeds Wildcats girls played in a tournament which included 350 other football teams from different parts of England.

Other small community organisations have also used sport to engage young British-Muslim girls and women. For example, the project 'Getaway girls' was set-up in 2005 to help young girls living in Leeds develop confidence and learn new skills in an environment that is supportive (GetAwayGirls 2009). In partnership with Leeds Metropolitan University, a project particularly aimed at South-Asian girls in Leeds, GetAwayGirls invited these young people to learn and play football for a week. However, this community development work is largely separate from the work of sports development agencies in the area and was a 'one-off' initiative rather than a long-term development. Unlike other urban areas in England that are home to large communities of British-Muslims, it becomes apparent that there is a lack in Leeds of programmes and initiatives to support sports involvement of young females.

The only one community organisation that does target young British-Asian female girls is Hamara that aims to foster healthy living within the local Pakistani community in South Leeds. Hamara, meaning 'ours' in Urdu, empowers young people directly to plan and develop youth projects through vehicles of sport, leisure and arts (Hamara 2009). On the basis of government funding to facilitate integration in Leeds, Hamara specifically uses football to 'bring communities together' (Hamara 2009). The West Riding County Football Association refers to the work of Hamara as evidence of good practice that arises when decisions are made jointly with various leisure providers that centralise voices, needs and experiences of young Muslim girls who wish to play the game (Hamada 2009).

Conclusion

In this chapter, pre-dominantly racist discourses about the problems of the hijab have been contextualised to provide an understanding of how British-Asian Muslim females in England grapple with such ideologies and material practices and, through their own forms of cultural resistance, search for a space of acceptance and belonging. Muslim women have, and are strongly developing, their own narratives and practices to legitimise their rights in general and specifically in relation to football. Their personal and political persuasions emerge in relation to their views about multicultural politics in England and their critical association to feminist politics. Some of the players reframe discourses about femininity to their own advantage and in addition some players re-work religious discourses and interpretations of the Koran to empower their involvement in sport.

References

Abbas, T. (Ed.) (2005). *Muslim Britain: Communities Under Pressure*. London and New York: Zed Books.

Ahmad, F. (2001). Modern Traditions? British Muslim Women and Academic Achievement. *Gender and Education*. 132: 137-52.

Ahmad, F. and Modood, T. (2003). *South Asian Women and Employment in Britain: the Interaction of Gender and Ethnicity*. London: Policy Studies Institute.

Ali, Y. (2000). Muslim Women and the Politics of Ethnicity and Culture in Northern England. In G. Saghal and N. Yuval-Davis (Eds.) *Refusing Holy Orders: Women and Fundamentalism in Britain*. London: Women Living Under Muslim Laws (MLUML): 106-30.

Asfar, H. (1994). Muslim Women in West Yorkshire: Growing up with Real and Imaginary Values amongst Conflicting Views of Self and Society. H. Asfar and M. Maynard (Eds.) *The Dynamics of 'Race' and Gender: Some Feminist Interventions*, London and Bristol: Taylor and Francis: 127-50.

Bagguley, P. and Hussain, Y. (2005). Flying the Flag for England? Citizenship, Religion and Cultural Identity among British Pakistani Muslims. In T. Abbas (Ed.) *Muslim Britain: Communities Under Pressure*. London and New York: Zed Books: 208-21.

BBC News. (2008). What is Your Game? Available at: http://news.bbc.co.uk/sport1/hi/football/your_game/7086786.stm [accessed: 14th May 2009].

BBC News. (2006). Straw's Veil Comment Sparks Anger. Available at: http://news.bbc.co.uk/2/hi/uk_news/politics/5410472.stm [accessed: 14 May 2009].

BBC News. (2005). Analysis: Segregated Britain? Available at: http://news.bbc.co.uk/1/hi/uk/4270010.stm [accessed: 14 May 2009].

BBC News. (2004). Headscarf Row Stops Soccer Match. Available at: http://news.bbc.co.uk/1/hi/world/asia-pacific/3662823.stm [accessed : 25 June 2008].

BBC News. (2002). Playing Football in a Headscarf. Available at: http://news.bbc.co.uk/1/hi/uk/1933713.stm [accessed: 14 May 2009].

Bhabha, H. (1994). *The Location of Culture*, London: Routledge.

Bonnett, A. (2000). *White Identities: Historical and International Perspectives*. Harlow, Prentice Hall.

Bowen, J. (2006). *Why the French don't like Headscarves*. Princeton: Princeton University Press.

Brah, A. (1996). *Cartographies of Diaspora: Contesting Identities*. London: Routledge.

Carrington, B. and Bramham, P. (2007). Disciplining Race: Leisure Studies and the Absent Politics of Race. Paper presented to the Leisure Studies Association Annual Conference , Eastbourne, UK, July 3-5.

Carrington, B. and McDonald, I. (eds) (2001). *Race, Sport and British Society.* London and New York: Routledge.

Cobb, C. (2007). FIFA Stops Short of Clear Ruling on Headscarves. [The Vancouver Sun Online] Available at: http://www2.canada.com/vancouversun/news/story.html?id=310889da-9f8b-4445-8ebb-24afb5291ed7&k=4682 [accessed at 14 May 2009].

Dagkas, S. and Benn, T.C. (2006). 'Young Muslim Women's Experiences of Islam and Physical Education in Greece and Britain: A Comparative Study', *Sport Education and Society*, 11 (1), 21-38.

Dwyer, C. (2000). Negotiating Diasporic Identities: Young British South-Asian Muslim Women. *Women's Studies International Forum*. 23 (4), 475-86.

Fortier, A-M. (2005). Pride Politics and Multicultural Citizenship. *Ethnic and Racial Studies* 28 (3): 559-78.

GetAwayGirls (2009). Home. Available at: http://www.getawaygirls.co.uk/index.html [accessed: 14th May 2009].

Giddens, A. and Hutton, W. (2000). (eds.) *On the Edge: Living with Global Capitalism*. London: Jonathan Cape.

Hall, S. (2000). Conclusion: The Multicultural Question. In B. Hesse (ed.) *Un/settled Multiculturalisms: Diasporas Entanglements, 'Transruptions'*. London: Zed Books: 209-41.

Hamara (2009). Who We Are. Available at: http://www.hamara.org.uk/hamara/about/ [accessed: 14th May 2009].

Hargreaves, J.A. (2004). Querying Sports Feminism: Personal or Political?' In R. Giulianotti (Ed.) *Sport and Modern Social Theorist*, Basingstoke. New York: Palgrave Macmillan: 187-206.

Hollows, J. (2000). *Feminism, Femininity and Popular Culture*. Manchester: Manchester University Press.

Kalra, V. (2000). *From Textile Mills to Taxi Ranks*. Ashford: Ashgate.

Lovell, T. (1991). Sport, Racism, and Young Women. In G. Jarvie (Ed.) *Sport, Racism and Ethnicity*. London; NY; Philadelphia: The Falmer Press.

McLoughlin, S. (2005). Mosques and Public Space: Conflict and Cooperation in Bradford. *Journal of Ethnic and Migration Studies*, 31 (6): 1045-66.

Miliband, D. (2009). 'War on Terror' was Wrong. *The Guardian* [Online 15 January] Available at http://www.guardian.co.uk/commentisfree/2009/jan/15/david-miliband-war-terror_[Accessed: 11 March 2009].

Modood, T. (2005). *Multicultural Politics: Racism, Ethnicity and Muslims in Britain*, Edinburgh: Edinburgh University Press.

Modood, T. (1997). Culture and Identity. In T. Modood and R. Berthould (Eds.) *Ethnic Minorities in Britain*. London: Policy Studies Institute: 290-338.

Office for National Statistics (ONS) (2001). Census Data. Available at http://www.ons.gov.uk/census/get-data/index.html [accessed: 15 May 2009].

Parekh, B. (2006). *Rethinking Multiculturalism: Cultural Diversity and Political Theory*. Second Edition, Hampshire; New York: Palgrave Macmillan.

Phillips, D. (2006). Parallel Lives? Challenging Discourse of British Muslim Self-Segregation. *Environment and Planning: Society and Space*, 24 (1): 25-40.

Phillips, D. Davis, C., and Ratcliffe, P. (2007). British-Asian Narratives of Urban Space. *Transnational Institute of British Geography*, 217-34.

Ratna, A. (2008). *British-Asian Females' Racialised and Gendered Experiences of Identity and Women's Football*. Unpublished PhD thesis: University of Brighton.

Raval, S. (1989). Gender, Leisure and Sport: a Case Study of Young People of South Asian Descent: a Response. *Leisure Studies* 8 (3): 237-40.

Runnymede Trust 2000. *The Future of Multi-ethnic Britain: The Parekh Report*. London: Profile and Runnymede Trust.

Scraton, S. (2001). Reconceptualising Race, Gender and Sport: The Contribution of Black Feminism. In B. Carrington and I. McDonald (Eds.) *'Race', Sport and British Society*. London and New York: Routledge: 170-87.

Scraton, S. (1994). The Changing World of Women and Leisure: 'Post-Feminism' and Leisure. *Leisure Studies*, 13 (4): 249-61.

Stillwell, J. and Phillips, D. (2006). Diversity and Change: Understanding the Ethnic Geographies of Leeds. *Journal of Ethnic and Migration Studies*, 32 (7), 1131-52.

Sporting Equals (2006). *Newsletter*, London: CRE.

The English Football Association (FA) (2009). Football for All. Available at: http://www.thefa.com/TheFA/WhatWeDo/Equality.aspx [accessed: 15th May 2009]

Turnbull, J. (2007). In the Islamic World, head scarves are not always compulsory. Available at: http://www.theglobalgame.com/blog/?p=242 [accessed: 25 June 2008].

Walseth, K. (2006). Young Muslim Women and Sport: The Impact of Identity Work. *Leisure Studies*, 25 (1): 75-94.

Watson, B. and Scraton, S. (1998). Gendered Cities: Women and Public Leisure Space in the Postmodern City. *Leisure Studies*, 17 (2): 123-37.

Werbner, P. (2000). Divided Loyalties, Empowered Citizenship? : Muslims in Britain. *Citizenship Studies*, 4 (3), 307-24.

Wilson, A. (2006). *Dreams, Questions and Struggles: South Asian Women in Britain*. London: Pluto Press.

Zaman, H. (1997). Islam, Well-being and Physical Activity: Perceptions of Muslim Young Women. In G. Clarke and B. Humberstone (Eds.) *Researching Women and Sport*. London: Macmillan Press Ltd: 50-56.

Chapter 10

Barcelona of the North? Reflections on Postmodern Leeds

Stephen Wagg and Peter Bramham

> Leeds is made up from many times and places. We need, what the Geographer
> Doreen Massey calls 'a global sense of space' to understand it more fully. Leeds
> is, and always has been , at once local and global, and everything in between....
> The past continues in the present, but what is often relayed is a version of the
> past which omits the story of everyday groups who sought to challenge certain
> developments and protect a different kind of present. Paul Chatterton (2006: 34)[1]

This book was inspired by, and has sought to analyse, the re-invention of the city
of Leeds during the late twentieth and early twenty-first centuries – its transition
from wool town and, latterly, the home of Don Revie's fractious footballers to
vibrant unofficial capital of a new North of England, a full-blown metropolis,
sprouting skyscrapers and boasting the first Harvey Nichols shop in the region.[2]
This final chapter reflects on the economic, political and social changes that have
befallen the city and discusses the apparent consequences of these changes for the
people who govern, promote, invest in or simply inhabit Leeds.

Postmodern Leeds: From Collectivism to Individualism

As was established in Chapter 1, Leeds is a largely de-industrialised space, its
economy now organised on predominantly post-Fordist principles. It aspires,
moreover, to be a postmodern city: that's to say, it depends economically on the
service sector – property development, office work and the various industries
devoted to culture, leisure and pleasure – and, as Janet Douglas observed in these
pages, there was in the mid-1990s a concerted attempt to turn it into a 'twenty four
hour city' on the European model. A small but significant part of the culture industry
that now thrives in and around a city like Leeds consists in now-eminent people
reflecting, via various media, on how life there used to be. Richard Hoggart and
Alan Bennett are clear examples here and more recently the novelist, playwright

---

1   Chatterton, P. 2006.Talking to the Past, Present and Future. A Critical Exploration
of the Many Faces of Leeds. *LSA Newsletter*, 74,July, 34-6.
2   This fabled store opened on the city's Victorian thoroughfare Briggate in 1996.

and academic Caryl Phillips (who was born in St Kitts, but grew up in Leeds) returned to the city to add his reflections.

Phillips typifies the relationship between local-made-made-good and the postmodern city, with the latter eager to embrace the cultural celebrity of the former: now Professor of English at Harvard University, he received honorary degrees from Leeds Metropolitan University in 1997[3] and the University of Leeds in 2003. His remembrance, on a visit in 2005, is predictable, yet evocative – telling of 'soot-blackened skies', of cobbled streets, stinking privies and red brick houses 'where people hung out their washing to dry as though vest and pants and bras were some form of celebratory bunting', his 'English playmates [who] were as thin and gaunt as their parents' and of his love for Leeds United, 'in their spotlessly white kit, a team who tormented opponents with industrial efficiency'. Now the flat caps, the belching chimneys and the 'Lowry-like figures' have departed and Leeds has become 'an architectural fusion of glass and metal' which 'civic leaders were now comparing to Barcelona. You had to be kidding me, right?' Softening, nevertheless, toward the shiny new Leeds, Phillips concedes that the waterside restaurants, hotels and apartment blocks that now define the city have helped to make it a 'newly vibrant area' – 'vibrant' being a virtually mandatory adjective in the lexicon of the global hawking of the postmodern city. But his concluding judgement is that

> the very flimsiness of this enterprise is clear in the reflected solidity of the buildings that sit all around. I walked through this new Leeds marvelling at change, but I also felt relieved at just how much of the old, including the dark, satanic, soot-blackened 50s, still remained.[4]

What of the 'flimsiness of the enterprise' to which Caryl Phillips refers? And, aside from the now sand-blasted-Victorian buildings in the city-centre, what remains of the solidities of the 1950s and before?

The fragility of the project that is the new Leeds is certainly not widely acknowledged in the public discourse of the city – largely because this discourse is invariably promotional and brooks only good news. It is aptly captured by the hollow business liturgy of 'civic boosterism' and 'city marketing'. Critical voices, however, have been raised: some of the strongest and most articulate emanating from the School of Geography at Leeds University. There Paul Chatterton and Stuart Hodkinson have developed a telling critique of Leeds the skyscraper city and the aims of its council to 'go up a league'. They have noted that a total of £10.4 billion was being spent on, or earmarked for, city-centre regeneration between 1997 and 2007 and that this principally entailed 'tall buildings' and 'booming

---

3   The racialised backdrop to the awards of honorary doctorates by Leeds Metropolitan University is explored in Ben Carrington's chapter in this book.

4   Caryl Phillips 'Northern soul' *The Guardian* 22nd October 2005. www.guardian. co.uk/artanddesign/2005/oct/22/photography.communities Access: 19th May 2009.

investment opportunities in high-value residential, office and retail property'.[5]
There are other major redevelopment projects across East Leeds, notably the
EASEL project covering the neighbourhoods of Halton Moor and Osmondthorpe,
Seacroft, Gipton, Harehills, Burmantofts, Richmond Hill and Lincoln Green. This
is the eastern part of the inner city to the North and South of York Road (the A64)
and includes many of the most socially deprived parts of Leeds. To this could
be added the renewal of housing in Swarcliffe, on the north-east outskirts of the
city, the regeneration of Little London, on the north edge of the city-centre and of
Beeston Hill and Holbeck, just south of the centre.

> What unites all of these initiatives is a regeneration mix in which refurbishment
> and remodelling of existing housing is combined with mass council housing
> demolition and sell offs to make way for new private housing.[6]

Chatterton and Hodkinson pointed to what has in any event been widely accepted
about the labour market in the postmodern city – that it is, in effect, two markets:
one offering high salaries to skilled workers and professionals in the 'knowledge
economy' and in financial and legal services and the other consisting of low paid/
short term/ part-time jobs in the service sector (catering, retail, tourism, leisure and
so on). They argue:

> Workers on the low road are actually the backbone of the Leeds economy
> – without them, the higher value jobs cannot function. The lack of affordable
> housing across the city and particularly in the more desirable green leafed
> suburban neighbourhoods threatens economic growth across the whole of the
> city. What we are seeing is the 'Brazilianisation' of the housing market, with key
> workers (increasingly migrants) commuting large distances from outlying towns
> such as Huddersfield and Dewsbury just to service, clean and run the essential
> components of Leeds' booming city-centre economy.[7]

The argument was clear: the poor (a term now largely abandoned in British politics)
and socially excluded (its favoured replacement), even if they had work in the
city, would stay poor. Moreover, would likely have nowhere within its boundaries
where they could afford to live. Leeds City Council had been one of the country's
largest providers of public housing and in 1979 had had around 100,000 council
houses let out at affordable rents to the working-class people of the city. In the
thirty years since then, constrained largely by the legislation of Thatcher and post-
Thatcher administrations, the city council has privatised or demolished around two

---

5   Paul Chatterton and Stuart Hodkinson 'Leeds: Skyscraper city' *The Yorkshire and
Humberside Regional Review* Spring 2007 pp. 30-31.

6   Chatterton and Hodkinson p. 32.

7   Stuart Hodkinson and Paul Chatterton 'Leeds: an affordable, viable, sustainable,
democratic city?' *The Yorkshire and Humberside Regional Review* Summer 2007 p. 25.

thirds of those. Not surprisingly, however, given the predominance of low-wage jobs in the local economy, demand for such housing has not diminished. For every council house newly available there were fifty candidates.[8] It's worth adding that in 2009 the *Yorkshire Evening Post* reported that Leeds had the greatest number of empty properties of any city outside London and that the bulk of these properties were at the top end of the market:

> An astonishing 17,557 properties are empty and not being used across the city, according to official government figures. The statistics are likely to infuriate the 24,444 households in Leeds who are currently waiting for social housing. The council house waiting list has grown by 4,800 during the last decade. The vast majority of the empty homes (15,297) are privately owned and many are believed to be executive one or two bedroom 'yuppie' flats, which were built during the city's boom years but are now un-sellable because of the credit crunch.[9]

Moreover, according to the same newspaper, the market for student accommodation in Leeds had reached saturation point two years earlier. Higher education is a major service industry in the postmodern city and it supplies other service industries such as rented property and various leisure outlets, such as bars, pubs and clubs: Leeds and Leeds Metropolitan universities between them bring well over 60,000 students into the city in any given year and the latter has also driven a number of the city-centre PFI building projects. In December of 2006, however, Planet Work Ltd (a company registered in Switzerland, specialising in introducing investors to openings in the property market), were refused permission to demolish a Victorian linen mill and replace it with student flats. There were various objections, among them those of Cllr Kabeer Hussain:[10]

> welcomed the decisions. "UNIPOL (a non-profit organisation providing student accommodation) tells us there is no need for yet more purpose-built student flats," he said. "This is an area which urgently needs family-friendly housing, including social housing. Perhaps developers will now listen".[11]

This may have proved a pious hope on the part of Cllr Hussain. Social housing in Leeds has continued to lag well behind demand and developers have generally not

---

8   Autonomous Geographies Pamphlet: 'Leeds Housing Policy and the Affordability Crisis' www.autonomousgeographies.org/leedshousingcrisis [Accessed: 18th May 2009].

9   '17,557 empty houses in Leeds' *Yorkshire Evening Post* 17th February 2009. http://yorkshireeveningpost.co.uk/news/17557-empty-houses-in-Leeds.4985208.jp   [Accessed: 18th May 2009].

10   Councillor Kabeer Hussain is the Liberal Democrat member for the Hyde Park and Woodhouse ward.

11   Howard Williamson 'Leeds hits 'saturation point' for student flats' *Yorkshire Evening Post* 2nd December 2009. http://yorkshireeveningpost.co.uk/news/Leeds-hits-39saturation-point-for.1910889.jp [Accessed: 18th May 2009].

seen it as a profitable proposition. At the same time the various PFI regeneration projects (an estimated 75 of these, worth over £4 billion, in the Yorkshire and Humber region were under way in 2006)[12] have brought lucrative business to the other sector of the Leeds job market. As in cities elsewhere, virtually all redevelopment schemes in Leeds are funded and administered through the controversial Private Finance Initiative. This covers all significant areas of city council activity including housing, education, health care and leisure provision. PFI is a global phenomenon, now practised in a range of countries across the world. It has been endorsed by the world's most powerful drivers of economic and social policy, the World Trade Organisation, the International Monetary Fund and the World Bank. It was inaugurated in Britain in 1992 by John Major's Conservative government but developed by 'New' Labour under Gordon Brown's Chancellorship (1997-2007), – despite a Labour Party Conference voting against PFIs in 2002. It is effectively a hire purchase agreement whereby various consortia finance public projects and are then repaid with an appreciable profit from public funds over a contracted period; on top of this in many cases the private corporations provide the services rendered by these institutions. PFI is regarded by many as *de facto* privatisation. Critics have argued, drawing from official data and filed Company Reports, that PFI projects are a 30 per cent more expensive than if government borrowed the money and kept the work in the public sector. Leeds hit national headlines in 2005 when one of its PFI hospitals at Seacroft was 'putting lives of staff and patients at risk'.[13] One Leeds councillor recently proclaimed the city the 'PFI capital of Europe' and, according to Claire Fauset of Corporate Watch in 2006:

> PFI contracts are predicted to bring in fees of £10 million for Leeds law firms this year alone. Lawyers DLA Piper have five partners specialising in PFI, each acting for projects totalling more than £1bn. The rest of the big six Leeds law firms are expected to have similar portfolios. DLA Piper have a contract with Leeds City Council to provide legal advice for all their PFI contracts.[14]

Leeds and Postmodern Politics: From Poverty to Social Exclusion

These commercial-political changes have of course been paralleled by changes in local and national politics. Central to these changes has been the re-casting of the policies and official philosophies of the British Labour Party in the wake of a third successive General Election defeat in 1987. The late 1980s and early

---

12   Claire Fauset 'Leeds: Live It, Lease It' *Corporate Watch Newsletter* 30 June/July 2006 www.corporatewatch.org.uk/?lid=2573 [Accessed: 18th May 2009].

13   Hencke, D.(2006) Private Finance Hospital putting lives at risk. *The Guardian* [On line 17.06.2005} Available at:www.guardian.co.uk [accessed :02.06.2009].

14   Claire Fauset 'Leeds: Live It, Lease It'.

1990s saw the rise of 'New Labour'.[15] The fundamentals of 'New Labour' are widely agreed. Labour strategists have come to accept market-based economics, globalisation (and the concomitant power of transnational corporations) and de-industrialisation. Since the decline of 'smokestack' and extractive industries has meant the withering of its historic bedrock support, the industrial working class, the party has, effectively, declared an end to 'class politics'. Instead it has given priority to equality of opportunity – between classes, gender groups, ethnicities and sexual orientations – and designated this new mission as the promotion of 'social inclusion'. At the same time there have been moves to marginalise trade unions, local party activists and the national party conference and to welcome corporate donors. In a widely quoted speech to California computer executives in 1998, Peter Mandelson, a chief architect of 'New Labour', reassured his audience that Labour was now 'intensely relaxed about people getting filthy rich'.[16]

The project of 'New' Labour had clear reverberations in the city politics of Leeds. Notably it was the scene of a show of strength and political intent by those seeking to re-direct the party. In 1997 the Labour Party in the Leeds North East parliamentary constituency adopted as their candidate Liz Davies, a London barrister, political activist and member of Islington Council. This diverse constituency encompassed affluent residential suburbs like Alwoodly and Moortown, each with large Jewish populations, up-and-coming neighbourhoods such as Chapel Allerton, home to many young professionals, and deprived inner-city areas like Chapeltown, the centre of Leeds' Afro-Caribbean community. It had been a safe Conservative seat for much of the post-war period and its MP from 1956 to 1987 had been Sir Keith Joseph, close political ally of Margaret Thatcher and scion of the wealthy Jewish family that ran Bovis the building firm. The Conservatives had held the seat once again in 1992 but the conversion of many large Victorian houses in the area to flats and multiple occupancy had seemed to make it winnable for Labour.[17] The candidature of Davies, perceived as precisely the kind of left-wing political activist the party now wished to be rid of, was rejected by Labour's National executive

---

15   See, for example, Mike Marqusee and Richard Heffernan *Defeat From the Jaws of Victory: Inside Kinnock's Labour Party.* London: Verso 1992; Stephen Driver and Luke Martell. *New Labour.* Cambridge: Polity Press 2006.

16   In a letter to *The Guardian* in 2008 Mandelson sought to stress the full context of this remark: 'You quote my comments to California computer executives in 1998 that "we are intensely relaxed about people getting filthy rich" (Leaders, January 11). I do not object to being quoted, as long as I am quoted accurately and in full. What I in fact said on that particular occasion was "as long as they pay their taxes". www.chickyog.net/2008/11/25/peter-mandelson-better-off-misquoted/ [Accessed: 29th May 2009]. It is, arguably, the part of the speech that has entered the public domain that matters, and not any qualification that Mandelson may have made at the time.

17   Seeukpollingreport.co.uk/guide/seat-profiles/leedsnortheast?cp=2   [Accessed: 29th May 2009].

Committee. Fabian Hamilton, a computer consultant, who had fought the seat for Labour in 1992, replaced Davies[18] and was elected to parliament in 1997.

After twelve years in office few, if any, vestiges of pre-1990 'Old Labour' politics remain. Fewer and fewer Labour politicians appear now to progress to parliament through the 'labour movement' (itself a virtually redundant term); many pass, at a comparatively young age, straight from university into politics, entering the Commons usually via jobs as political researchers or 'heads of policy'. There have been some tangential ironies here for those active in socialist politics in Leeds. Hilary Benn, cabinet minister and Labour MP for Leeds Central since 1999, a former researcher at the 'white collar' union ASTMS and Head of Policy for Manufacturing Science and Finance, is the son of Tony Benn, longstanding tribune of the parliamentary left and the man seen by 'New Labour' politicians as rendering Labour unelectable in the 1980s. Fellow cabinet ministers and 'New Labour' strategists David and Ed Miliband, who have spent all their working lives as political researchers, analysts or policy makers, are the sons of Marxist intellectual Ralph Miliband (1924-94) who was Professor of Politics at Leeds University between 1972 and 1976 and one of the Labour Party's sternest socialist critics.[19] Ed became MP for nearby Doncaster in 2005; two years later he was made a cabinet minister, still aged only 38.

A debate in the House of Commons in 2003 provided an excellent vignette in which to observe the recent changes in the politics of Leeds, of parliament and of their intersection. On the evening of 14th October that year, George Mudie, Labour member for Leeds East, rose to speak to honourable members. Mudie, by then 58, had been in parliament since 1992. He was a political veteran having been on Leeds Council since 1971 and, irrespective of his political proclivities, his route to parliament had been more traditional than in many contemporary Labour Party biographies: he is a former engineer and merchant seaman and was an official of the National Union of Public Employees in the late 1960s. Leeds East encompasses some of the poorest parts of the city, including Seacroft, Gipton and Harehills. Mudie told the House that 25% of households in Leeds now claimed some kind of mean-tested council benefits and that that figure rose to 40% for some inner-city districts.[20] These same districts showed high rates of heart disease, teenage conception, poor educational achievement and domestic burglary. While, historically, the estates which he represented had delivered working-class people from inner-city slums, they were now seventy years old and run down. Most

---

18   Davies, based at Garden Court Chambers in London, is a specialist in housing law; she regularly represents the homeless and travellers see www.gardencourtchambers. co.uk/barristers/liz_davies.cfm [Accessed: 29th May 2009].

19   See in particular his *Parliamentary Socialism: A Study in the Politics of Labour* London: Allen and Unwin 1961.

20   I have quoted at length, both directly and indirectly, from this debate. All quotations are taken from Hansard Inner City Poverty (Leeds) HC Deb 14 October 2003 Vol.411 cc80-96, available at hansard.millbanksystems.com/commons/2003/oct/14/inner-city-poverty-leeds [Accessed: 13th May 2009].

of the housing needed modernising or replacing. However, Mudie lamented, the government had obliged the city council to place its housing policy under the control of six 'ALMOs' (Arms Length Management Organisations). These ALMOs had then been inspected, found to be inadequate and, thus, denied money for capital spending. Colin Burgon, Labour MP for neighbouring Elmet, on Leeds' largely rural eastern periphery, intervened to offer his support. He'd been born in Gipton, he told the House, and taught for many years in Seacroft: at that time there had been community cohesion in these districts but this was now threatened because of the loss of the engineering and tailoring industries. The *Yorkshire Evening Post*, Mudie continued, were running a 'Life in Leeds' campaign and describing his constituents as the *People A City Forgot*, 'their lives blighted by fear, violence, crime and poverty'. Unemployment was rising and, added Mudie, many jobs currently available in Leeds would likely not last another decade:

> I do not believe that 25,000 people will be working in call centres in 10 years' time. Voice chips will have replaced those jobs and people currently in low-paid work in our inner cities will be looking for alternative employment.

Citing a recent government booklet bearing the signature of Prime Minister Tony Blair, Mudie was withering: 'We cannot feed the hungry with statistics on national prosperity'.

The parliamentary response to George Mudie was principally twofold and telling. First, fellow Labour MP Fabian Hamilton (Leeds North East) rose to acknowledge deprivation in the Chapeltown area of his constituency but preferred to define difficulties there in the language of public order:

> The problem involves crime related to crack cocaine, heroin and street gangs that use firearms and the most appalling violence against one another, but innocent victims get in the way.

Would Mudie support him in calling for a 'crackdown on crime'? Mudie's reply, like much of his exposition, came out of a politics now perceived as anachronistic in the Labour Party:

> We have enough sticks and we are adding more sticks to the armoury, but I wish there were some carrots. I look at some of the kids on the estates and I cannot defend their behaviour. I attack their behaviour and I want them to be dealt with, but I look at the houses they come from, the peer pressure and the parents' behaviour, and I worry for the kids.

For Mudie, then, there was still some qualified space for the view that social factors informed and influenced social behaviour. Increasingly, in the self-transforming Labour Party, such notions were inadmissible.

The second, more official response to Mudie's plea for assistance for the people of Leeds East came from the government in the person of Yvette Cooper, Parliamentary-Under Secretary of State in the Office of the Deputy Prime Minister. Cooper, then 34, had had a typically 'New Labour' trajectory. The daughter of an official of Prospect (a trade union for engineers, managers and scientists) and a teacher, she had been educated at the universities of Oxford and Harvard and worked as a researcher for the previous Labour leader, John Smith, and as economics correspondent of *The Independent* before entering parliament in the Labour landslide of 1997, as Member for Pontefract and Castleford – a few miles from Leeds. In her reply Cooper offered warm words (both to Mudie and to the new Leeds) along with assurances that analyses were being undertaken, proposals developed and discussions begun. 'My Honourable Friend' she conceded:

> gave an eloquent account of the lives that his constituents lead. He mentioned housing on estates, the services that people receive, the opportunities they have and their sense of community, including whether they feel safe on the streets. We need to recognise that Leeds faces huge challenges. It is a city of great contrasts and inequalities. In many ways, Leeds is an amazing city. It is a powerhouse for economic growth in the region. My constituency is benefiting substantially from being close to Leeds and its economic growth. Considerable regeneration is taking place. It is clear when one drives through the centre that the skyline is crammed with cranes, such is the building and investment in the city.

Burmantofts, Harehills and Seacroft, she acknowledged, were among the 10% most deprived wards in the country: there might be funding for neighbourhood wardens in some of these districts. The Social Exclusion Unit (part of the Office of the Deputy Prime Minister) was 'carrying out a detailed analysis on jobs and enterprise'. The West Yorkshire Partnership was 'developing proposals for the regional housing board to fund areas such as Beeston and Harehills'. 'My personal view', she asserted toward the end of her remarks' 'is that the programme that is doing most to tackle inner-city poverty and inequality is Sure Start'.

There are a number of 'New Labour' signifiers here: emollient talk; recourse to units, discussions, strategies and the like; rumours of proposals; intimations of social control as a priority (the wardens); and, perhaps most crucially, the citing of Sure Start as the major boon to the urban poor. Sure Start is a programme of child care, health and pre-school education modelled on President Johnson's Project Head Start, inaugurated in the United States in 1965. It has an inherently *individualistic* basis, in that it does not offer material help to communities in the here and now; it offers a start in life – and, thus, a chance of escape – to the small children of these communities, albeit many of them deprived.[21] The politics of

---

21   For details of Sure Start see Stephanie Northern 'Sure Start' *TES Magazine* 28th January 2005 Available at www.tes.co.uk/article.aspx?storycode=2069291 Access: 29th May 2009.

'New Labour' conceived in London, but reverberating in Leeds and dozens of British towns and cities likewise, were loosely postmodern. They embodied the belief widespread among postmodern theorists that the notion of social progress was a flawed 'meta narrative'; they rejected cradle-to-grave welfarism, and concerned themselves, via such schemes as Sure Start, principally with the cradle. They placed social difference ahead of social class and rejected the 'masculinist collectivism'[22] of Labour's trade union past. The tired collectivism of a 'one size fits all' public sector had to be modernised. Service delivery must henceforth be focused on individual consumers who now had to make the right lifestyle choices in education, health and housing. 'New Labour' policies also placed a high value on impression management: sound bites, rhetorical flourishes, press releases, photo opportunities, strategies, spin doctors and the like.

The contention of Leeds-based Polish social theorist Zygmunt Bauman (1992)[23], is that the majority of the population in post-modern times are seduced into consumer culture; the excluded minority (i.e. the poor, unemployed and racial minorities) yearn to consume too. The nation state is no longer interested in seeking legitimation, in binding producers into work, in socialising citizens into a homogenous national culture. Both central and local politicians have relinquished the task of social integration to the market and the media. People are engaged in society as consumers not producers: spending therefore becomes a duty and political partisanship an irrelevance. The current neo-liberal hegemonic project is to produce willing consumers rather than obedient citizens immersed in national culture. Market inequalities may produce flawed consumers – the poor, homeless and unemployed but they are in their turn subjected to disciplinary power, Foucauldian surveillance and control. The bottom third in postmodern society are left behind , rooted in local working-class communities, as footloose global capital restlessly searches for new places for investment.

It is therefore possible that this parliamentary debate passed unnoticed in Harehills, Burmantofts and Seacroft but, when all the fine words had been uttered, George Mudie had been given little tangible to offer the people who lived there. The Labour Party as a vestigial organ of democratic accountability, through which grievances might be expressed and the occasional wrong righted, had largely disappeared. On this point, Hodkinson and Chatterton are pessimistic. The growth of PFI, they argue, will make it more and more difficult to determine who is responsible for what:

---

22   A phrase employed, for example, by Ian Taylor, Karen Evans and Penny Fraser in *A Tale of Two Cities: A Study in Manchester and Sheffield.* London: Routledge 1996, p. 23.

23   See Bauman Z.1992. *Intimations of Postmodernity.* London: Routledge. Professor Bauman is a retired Professor of Sociology from Leeds University who is much quoted on issues relating to individualism, postmodernity and liquid modernity. Although writing on the postmodern, he himself has become a local celebrity intellectual, akin to Anthony Giddens, one time director of the LSE and key intellectual luminary for Blair's 'Third Way' as political 'New Labour' solution to globalisation.

Ask a school teacher about changing a light bulb in a PFI school, or try to track down who exactly is now running your PFI council housing, and you will soon realise local accountability for local services is a thing of the past. The more that public services and assets are contracted out or sold off to global companies, the less direct power our local representatives have to respond to the urgent problems of the day – such as climate change, social inequality or workers' rights.

Leeds City Council is a major holder of land assets and, in order to finance the regeneration, these assets will be sold or given away to encourage private developers to take on the works. Most of the city-centre is already in the hands of the private sector, which is rolling out Business Improvement Districts to manage and develop their patch. This threatens democratic control and a loss of future sovereignty. We can see exactly what kind of democracy we get in the 'skyscraper city' by the bogus 'consultations' that accompany these regeneration schemes. The loss of public assets and public space to an increasingly corporate-owned city only increases the power of big business and decreases the autonomy of people to build their own futures. Meanwhile, few councillors are prepared to make a stand on issues of local accountability.[24]

Nevertheless, an axis of local journalists and lecturers from the city's universities have continued to press the issue of accountability and in March of 2008 the University of Leeds staged a forum entitled 'Leeds: Are We Going in the Right Direction?' and its theme was the gentrification of Leeds – that is, the forcing out of the city of its poorer people.[25] The *Yorkshire Evening Post*, whose recent investigations were cited as an inspiration for the conference, reported:

Key topics raised at the meeting, chaired by Andrew Edwards of BBC Radio Leeds, included complaints about lack of affordable housing and transport issues. There was praise for the historic architecture, the quality of independent shops, pubs and cafes and the city-centre's compact, "walkable" nature. But there were worries that Kirkgate Market – once regarded as a jewel in the city's crown – was not promoted properly and was losing its appeal; plus grievances about the failure to utilise riverside land as public areas; lack of green space and trees; lack of evening activities that did not involve drinking; and failure to properly consult people on important changes ....[26]

---

24 Hodkinson and Chatterton 'Leeds: An affordable ....' p. 26.

25 A transcript of this debate can be found at: www.threshfield-uarry.org/groups/leedsdirection/debatefeb2808.doc.

26 Debbie Leigh 'Future of Leeds: "People want a say in how their city's run"' Yorkshire Evening Post 3rd March 2008. Available at www.yorkshireeveningpost.co.uk/news/Future-of-Leeds-39People-want.3834692.jp [Accessed: 30th May 2009].

This appeared to cut little ice with Leeds Council Leader Andrew Carter, the councillor with responsibility for development and regeneration, whose response was: 'Nobody much benefits from unsubstantiated criticisms thrown from the academic sidelines'.[27]

Fragmented Leeds: Living below the Bottom Line

Councillor Carter notwithstanding, academics and media journalists have provided the best indications of what life might be like in the impoverished districts of inner-city Leeds. Some districts have acquired - indeed have for some time been saddled with – dismal reputations. Moreover, while sociologists and others talk of social conditions, others take the increasingly individualistic political discourse of the twenty-first century at face value: for them these areas have become degenerated simply because they are inhabited by degenerates. These districts are inhabited in large part by the people who, as Chatterton and Hodkinson put it, take 'the low road' in the labour market and analysis of the post-Fordist global economy suggests that they may constitute up to two thirds of the labour market as a whole.[28]

The region's leading evening paper recently alleged that the city had forgotten them. Certainly, as we have argued, the families of these districts can now expect little material support from the Labour Party, which now represents what the journalist John Pilger has called a 'single-ideology business state'.[29] Indeed Zygmunt Bauman has argued that there is no longer a basis for distinguishing between right and left in politics and that the state no longer undertakes to safeguard 'existential security'. A fundament of the now-obliterated left, he argued,

> was to believe that it is the sacrosanct duty of community to care for and to assist all its members, collectively, against the powerful forces they are unable to fight alone.[30]

---

27   ibid.

28   See for example Will Hutton (1995) *The State We Are In.* London: Jonathan Cape. He argues we now live in a '40-30-30' society – the rich and the poor have now been separated by an insecure middle class made up of casual, flexible or part-time workers, usually women. For a full discussion of changing class structures and of Ulrich Beck's thesis of individualisation see Chas Critcher, and Peter Bramham. The Devil Still Makes Work in Haworth and Veal (eds.) *Work and Leisure.* London and New York: Routledge 2004, pp. 34-50.

29   John Pilger 'The depth of corruption' *New Statesman* 28th May 2009 Available at: www.newstatesman.com/uk-politics/2009/06/pilger-blair-iraq-british-mps [Accessed: 8th June 2009].

30   Zygmunt Bauman. A political muddle *The Guardian* 26th October 2008. Available at: www.guardian.co.uk/commentisfree/2008/oct/26/labour-economics [Accessed: 8th June 2009].

Responses in forgotten Leeds to these new political and economic circumstances have ranged from violent protest, through appeals, polite or impassioned, for a change in council policy, and conventional political campaigning to the exploration of alternative political arrangements and ideologies. This section offers a glimpse of each one.

When Caryl Phillips visited Leeds in 2005 he noted that a great many of his beloved redbrick back-to-back houses were now:

> home to Leeds' Muslim population, many of whom, as we know from the events of July 7, remain unconvinced by civic proclamations of racial and ethnic harmony. Three of the four suicide bombers who attacked London were from the greater Leeds area and, although they were second- or more properly third-generation citizens of Leeds, their disaffection with Britain clearly ran deep.[31]

Indeed there had been strong evidence of this disaffection four years earlier when the Harehills area was the scene of violent disturbances. An analysis of these disturbances, which predictably ran counter to most official versions, was produced the following year by Max Farrar, a sociologist at Leeds Metropolitan University and author of an earlier study of Chapeltown.[32]

Farrar noted the frequent recourse to particular words in media descriptions of these events. The *Yorkshire Evening Post*, for instance, called the incident a 'riot', perpetrated by 'outsiders'.[33] These words rendered the disturbances doubly illegitimate: the perpetrators were said to come from outside (implying the causes did not lie in Harehills itself) and they constituted a 'riot' - that is, they were senseless, since 'riots' can never be justified. Indeed, the local MP, Fabian Hamilton, was concerned to define the incident solely in terms of individual malevolence, permitting it no social dimension. In an email to Farrar, he said: 'I think the arrest [of a local man] was used as an excuse by very angry/wicked elements in the Asian community. Somebody wanted a riot and was determined to lure the police into the area, which they did'.[34] Renditons of this kind are characteristic of contemporary political discourse, in that they do not allow for the possibility of *explanation* – beyond the tautological assertion that bad acts were committed by bad people simply because they wanted to commit them. For many in public life explanations belong on Councillor Carter's 'academic sidelines'.

Farrar suggests that Harehills is a 'space of representation' – that is, the name represents something in Leeds: in this case, it's seen as the part of the city where

---

31   Caryl Phillips. 2005 'Northern soul' The Guardian Colour Supplement p. 26.

32   Max Farrar *The Struggle for 'Community' in a British Multi-Ethnic Inner-City Area* Ceredigion: Edwin Mellen Press 2002.

33   6th June 2001, p.1 Quoted in Max Farrar 'Parallel Lives and Polarisation' British Sociological Association 'Race' and Ethnicity Study Group Seminar, City University London, 8th May 2002 p. 3. Available at www.maxfarrar.org.uk/docs/HarehillsBSARace2May02.doc

34   Farrar p. 14.

low-income British Asians live (rather as the word 'Gipton' signifies the low-income 'whites' of Leeds). They live there because it offers the cheapest and oldest property in the city and because they are racially excluded from other areas. Paradoxically, this recognition explains why most of those convicted for the disturbances of 2001 came from other parts of town: they got mobile 'phone messages that something was 'kicking off' and came over to the 'symbolic space' of 'Harehills' to show solidarity. The incident was apparently triggered by the arrest of a 31-year-old British-Bangladeshi male on suspicion of theft: CS gas was said to have been used in his arrest and protests had mounted. A local youth worker suggested to Farrar that the police routinely harassed young Asians of Harehills for doing things that might have be thought comparatively harmless when done by other young people – sitting in their cars, driving around, 'chilling out'. 'The majority just smoke weed', he said. Young Harehills residents therefore were, in this account, being racialised – singled out for their supposed 'race'.[35] Official spokespeople insisted that there was no racial dimension to the events of 2001; to have said otherwise would have brought the whole affair into an area of acknowledged public responsibility – social exclusion and anti-racism.

Farrar also argued that masculinity had played its part in the protests. The predominantly Pakistani- and Bangladeshi-British lads of Harehills wanted to show the police that they were not to be messed with. They wanted to control their 'turf' against the police – and anyone else (Caribbean youth, for instance) who had thought they might be a soft touch. There may also have been a 'carnivalesque' element to the protest – a sort of jubilation through creating temporary mayhem.

This, then, was an explanation. But it was not necessarily a comforting one for those seeking evidence of radical politics or insurrection. When a number of arrests were made no defence committee had formed in Harehills and when a battery of prison sentences was handed down, no significant protest registered in the area. Farrar noted a 'segmentation' by generation – a section of local youth culturally adrift from their parents.[36] Despite popular fears, Muslim 'fundamentalism' did not seem to feature among the cultural flashpoints in this community; hard drugs, altercations with the police and petty crime did.

Ultimately, *all* named spaces are likely to be spaces of representation. While districts like Harehills have, in contemporary discourse about the post-modern city, been rendered as self-segregating spaces, brooding and incubating religious extremism (see Aarti Ratna's chapter in this book), 'white' districts have been popularly depicted as home to a degenerate underclass, made up of 'dysfunctional' families and 'chavs'. An affectionate example in contemporary popular culture is Channel Four's comedy drama *Shameless* (2004-), the ongoing story of the amoral, conniving, but ultimately lovable Gallagher family, described

---

35   Farrar 'Parallel Lives...' pp. 10-12. The word 'race' now always appears in inverted commas in sociological discourse. It is seen simply as something which some people impute to others but as having no scientific validity as a way of categorising humans.

36   Farrar 'Parallel Lives...' p. 8, 13-16.

as 'dysfunctional' on the programme's own website.[37] However, when in 2008 a woman on a council estate in Dewsbury, ten miles from the centre of Leeds and home to many of the city's 'low road' workers, replicated a scam earlier featured in an episode of *Shameless*, the national press deployed its vocabulary of underclass and dysfunction with considerable animus. Karen Matthews faked the disappearance of one of her daughters in order to claim reward money for her recovery. At the time of her subsequent trial the *Daily Telegraph* wrote of Matthews' 'dysfunctional family' and 'feckless life' which 'revolved around beer, pizzas and her 60-a-day cigarette habit, funded by benefit payments of £286.60 per week which were boosted every time she had another child'.[38]

There are now websites to accommodate this sort of scorn for the underclass; one is called 'Chavtowns: The Nemesis of Estate Agents and Local Councilors [sic]'. 'Chavtowns' consists of self-parodic rants that reduce the residents of variously despised districts to cartoon characters reminiscent of the comic *Viz*. This, for example, purports to describe the Beeston area of south Leeds:

> Grim doesn't even come close to describing this insalubrious suburb tucked nicely inside Leeds' anal sphincter. Dewsbury Road, the areas filthy main thoroughfare, cuts a sweeping arc through scenes of urban decay that wouldn't look out of place in inner city Detroit. Dirty shellsuits and grubby underpants hang across streets of soot covered back to backs, while teenage tearaways aboard stolen Piaggios hare like crazed maniacs through sink estates [......] A short wheel spin away down in the more affluent end of Beeston lies that most venerable of Leeds institutions, The Tommy Wass public house. On any night of the week, you're guaranteed to see the spectacle of 6 or so, hoopy earringed teenaged trollops staked outside the main doors of the said bar or by the traffic lights on the corner of Old Lane. It's in these locations that these fat arsed clamydia [sic] ridden bints accost bewildered males on their way to the local Spar for a 6 pack. In this instance, refusal always offends and the male in question whether 17 or 70 will invariably be greeted with a blizzard of obsenties [sic] worthy of a whole platoon of troopers.[39]

Another describes Gipton as 'scummy, chav infested shitheap of epic proportions'.[40] Dewsbury, Beeston and Gipton thus become spaces of grotesque representation, with two important consequences. First, poverty and neighbourhood decay are

---

37   www.channel4.com/programmes/shameless [Accessed: 8th June 2009].

38   Paul Stokes 'Shannon Matthews trial: The dysfunctional family where children equalled benefits' *Daily Telegraph* 4th December 2008. Available at: www.telegraph.co.uk/news/newstopics/politics/.../Shannon-Matthews-trial-The-dysfunctional-family-where-children-equalled-benefits.ht... [Accessed: 8th June 2009].

39   www.chavtowns.co.uk/modules.php?name=News&file=article&sid=828 [Accessed: 1st June 2009].

40   www.chavtowns.co.uk/2004/11/gipton-leeds/ [Accessed: 1st June 2009].

increasingly individualised, rendered as the fault of the local residents – the females among them styled as 'feckless', over fecund 'trollops', undeserving of sympathy or state aid.[41] Second, on the rare occasions when people who live in these districts are themselves permitted to describe what life is like there, they have to write against these widely-canvassed stereotypes of their place. An example of this can be found on the BBC website, which contains a section entitled 'Where I Live'. Pages within this section, called 'Mini Guides', carry accounts of life in individual city districts written by people who live there. Recently there were pages on Holbeck (posted in 2006) and Gipton (2007). The writers have written powerfully, and variously, of a lost sense of community in their area and have made polite appeals to Leeds Council for assistance in retrieving it. Nighat Qureshi, who'd lived in Holbeck since 1994 when he was 17, wrote:

> Holbeck has got a bit of a bad reputation. There are some bad areas which are full of crime as well as prostitution. Two areas of Holbeck are full of prostitutes. Women living in the area are quite intimidated. They don't feel comfortable walking down the streets. [....] Drug related incidents are an everyday phenomen[on]. A few times I have seen people selling drugs in the streets, although I don't know them personally. Years ago Holbeck used to be a posh area, it wasn't full of vandalism or crime. According to some people Holbeck is a dump, but in my opinion it is not a dump at all. It just needs a lot of help to eliminate all the crime from the streets of Holbeck. Holbeck is a multi-cultural community. The part of Holbeck I am living in is very pleasant. Everyone knows each other and they are all very friendly with each other as well.[42]

Audrey Busfield, while acknowledging drugs, prostitution and break-ins, was not so sure about the community spirit:

> There is no unity now like there used to be. No community anymore. It is a shame because it could be made a nice place to live in if things were sorted out.

Another resident responded: 'I used to live in Holbeck, but moved to Beeston, not a very good move. It's time Leeds City Council looked after all it's citizens, not just a few'.[43] On another page 35 year-old Kara Marsh, who had spent all her life in Gipton, wrote in similar, but politically more contradictory, terms:

---

41   Something acknowledged, bizarrely enough, by the *Daily Telegraph*. See Neil Tweedie. Another side to Shannon Matthews's Moorside. *Daily Telegraph* 27th March 2008. Available at www.telegraph.co.uk/news/features/3636060/Another-side-to-Shannon-Matthewss-Moorside.html [Accessed: 14th May 2009].

42   www.bbc.co.uk/leeds/content/articles/2006/05/26/mini_guide_holbeck_feature.shtml [Accessed: 1st June 2009].

43   ibid.

I grew up on the St. Wilfrids [estate]. When I was younger the area was so much prettier than it is now. The streets were made up of semi-detached houses and large gardens bordered by perfectly trimmed green hedges. We had colour, in that people used to keep their gardens well-maintained with a wide range of flowers and bedding plants. There would be an occasional hanging basket too, dripping with the vibrancy of summer. [....] We had trees that produced pretty pink and white blossom, which fell like confetti to litter the path. Children would collect it by the handful and attempt to make perfume. Of course we've always had 'dysfunctional' families but on the whole these were few in number. Most families were just unfortunate in that jobs became scarce. All my uncles, aunts and elders tell me, you could walk out of one job and straight into another. [....]

The writer explicitly blames economic factors – inadmissible in most political discourse on local poverty:

All I ever hear around our way is "I can't afford it". Ever wonder why there's a big debt problem in this country – IT'S BECAUSE THE COST OF LIVING IS TOO HIGH! Wonder why there is a massive epidemic of family breakdown – ITS BECAUSE THE COST OF LIVING IS TOO HIGH and that's whether families are working or not.
Gipton estate looks awful, which is probably the reason the council has proposed they want to knock the whole place down....[....][44]

As with the other accounts, the writer lamented the loss of decent, tidy, community-spirited living in her district and castigated the local authority ('and all the pinstripes in this country') for letting it decay:

Blame the state of things on dysfunction if you like but if you're honest you'll be aware that the majority function very well, thank you very much – despite the circumstances!

Although Kara Marsh rejects racism – 'None of us want that' – she complains of 'too many different groups of people' on her estate and argues that

If you're new to the UK, there are lots of initiatives but what if you were born here, what about our own and what about our youth? What's happened to their opportunities?

These remarks are politically incoherent and they can be read as an expression of despair with the main parties and probably not untypical among Gipton residents.

---

44   www.bbc.co.uk/leeds/content/articles/2007/08/17/civic_life_kara_gipton_feature.shtml [Accessed: 1st June 2009].

They also mark a susceptibility to far right voting. While the neo-fascist British National Party won no seats on Leeds Council in the local elections of 2007, and the Harehills and Gipton ward was held by the Liberal Democrats, in the European elections of 2009 the BNP won two seats, – in the North West and Yorkshire and Humber, a broad area with Leeds at its heart. Moreover a poll by Channel Four News found that seven out of 10 BNP voters (and almost as many Green Party and United Kingdom Independence Party voters) thought that there was 'no real difference these between Britain's three main parties' and that 59% of BNP voters thought that [the] Labour [Party] 'used to care about the concerns of people like me but doesn't nowadays'.[45]

Little London lies to the north of inner-city Leeds and consists largely of high-rise flats and maisonettes built by the council in the 1960s. In November of 2006 it was still home to 1,400 council tenants and the government approved plans to spend £95 million on improving the estate, once again on the basis of a PFI agreement. According to Stuart Hodkinson's account of the ensuing controversy, the understanding of residents was that their homes would be brought up to government standards of decency and affordable private housing.[46] When it emerged that 300 flats were actually due to be vacated and sold to a private developer an unsuccessful challenge was brought in the High Court. Tenants accused the government of deliberately running the flats down and denying money to the ALMOs concerned – a general issue raised in parliament by George Mudie. The result was a Save Little London Campaign, one whose members called attention to

> BMW-driving professionals in £1,000,000-plus private homes on re-branded areas of the estate visibly gated from the rest of the area, while tenants have had their own homes turned into building sites for nearly two years .... [47]

This was defined by the council as a 'sustainable mixed community' and by its opponents as 'gentrification' – what Kara Marsh had perhaps perceived as compulsory mixing of *ethnic groups* is here seen, more accurately, as the encroachment of one *social class* upon another. Indeed a local consultancy firm called *Outside* had called for housing in Little London appropriate 'especially to young professionals who are currently priced out of the market'. Hodkinson's article was accompanied by the Council's dissent from its arguments. 'The PFI scheme', it said, 'is the only way that Little London is going to get investment

45   Channel Four News 'Who voted BNP and why' 8th June 2009. www.channel4.com/news/articles/politics/domestic_politics/who+voted+bnp+and+why/3200557 [Accessed 8th June 2009].

46   Stuart Hodkinson 'Little London Takes the Initiative' *Big Issue in the North,* 16th-22nd April 2007 p.14 Available at: www.autonomousgeographies.org/files/bissue_apr07_p14.pdf [Accessed: 18th May 2009].

47   Hodkinson 'Little London.....' pp. 15-16.

on this scale'.[48] The following year, however, the campaign to have the flats refurbished for their existing tenants appeared to be vindicated when £2.3 millions were made available from the government's Decent Homes fund for this purpose. Councillor Les Carter, responsible for Housing, said:

> I am pleased that we have been able to take on board the residents' views following the consultation, and I am sure that the people of Little London will notice very real improvements to their neighbourhood when work gets under way.

The Campaign now felt entitled to question the reason they had been given for the proposed privatisation – that 'the Lovells' (the Lovell Park Grange apartment block) was no longer viable.[49]

The campaign to salvage 'the Lovells' for council tenancy, which had supportive coverage from the *Yorkshire Evening Post* and local television, is probably the most successful attempt to confront the regeneration of Leeds by the contemporary alliance of Leeds Council and local and/or global capital. It used conventional pressure group tactics and they worked. Hodkinson and Chatterton, though, have acknowledged the existence of a culturally deeper form of protest in the form of the social centres movement – a series of 'autonomous spaces' scattered across the UK, dedicated to anti-capitalist philosophies and activities and represented in Leeds by The Common Place, situated on Wharf Street, near the city-centre. As Hodkinson and Chatterton point out, these centres face many problems – they are often run, for example, by middle-class people who have day jobs – and are usually short-lived. The Common Place has also had the difficulty of functioning within the capitalist system – unlike its predecessors A-Spire (1999-2005) and Maelstrom (2005) which occupied empty buildings, Common Place is a rented space.[50] In July 2008 Leeds Council withdrew The Common Place's entertainment licence – its principal source of income. The council's case was that the Common Place was not a *bona fide* members club – a condition of the licence – and was lax about signing non-members in. The suspicion was, however, that the council did not want to sanction a city-centre base from which opposition to its core policies might be organised: Carl Gallagher of Leeds law firm Zermansky and Partners, who acted for The Common Place, calculated that Leeds City Council had spent over £50,000 in bringing their case to court in a attempt to shut down the centre.[51]

---

48   Hodkinson 'Little London ….' p. 16

49   Save Little London Campaign website: savelittlelondon.blogspot.com/ Access: 1st June 2009.

50   See Stuart Hodkinson and Paul Chatterton 'Autonomy in the city? Reflections on the social centres movement in the UK' *City* Vol.10 No.3 December 2006 pp. 305-15. Available at www.autonomousgeographies.org/publications [Accessed: 1st June 2009].

51   The Common Place website: www.thecommonplace.org.uk/ [Accessed: 1st June 2009].

Conclusion: Postmodern Leeds – Living with Contradictions

The central contradiction for the socially excluded of Leeds – as for similarly placed people in so many comparable cities across the world is that, on the one hand decision makers in the city need a reservoir of cheap labour with which to attract firms (a new supermarket, perhaps, or a call centre) while, on the other hand, practically the only political assistance to the working class is, in effect, is the slim chance for individuals to escape, via pre-school booster programmes. Capitalists, quite simply, are trawling Leeds in search of profit. There was therefore something deeply symbolic in a remark made by George Mudie on the House of Commons in February of 2008: 'On the motorways in the Leeds area in the morning and evening, one can see the number of people who take employment in Leeds and then spend their money elsewhere'.[52] Global capitalism does not entail the principle of 'putting something back'.

Those who run Leeds must continue to find ways of attracting capital and visitors to the city and slogans concocted by highly paid consultants of impression management will be part of this process. Hence the claim that Leeds is now the 'Barcelona of the North'. But Barcelona, no less than Harehills, is a 'space of representation' and those who invoke its name wish to conjure up images of a romantic Iberian city, with attractive restaurants and hotels and a dazzling football team. But Barcelona, like every other city, has its slums – the Southern Raval district, for example, is, in effect, Barcelona's Gipton. Leeds publicists will of course continue to maintain the city as a space of seductive representation and there are some important recent examples of this. In 2006 The Who returned to Leeds University to reprise their concert there of February, 1970 which resulted in the album 'The Who Live at Leeds'. A blue plaque was unveiled and the Vice Chancellor, Prof. Michael Arthur, reasserted the lack of distinction between high and popular culture in the postmodern era:

> Leeds has had its Nobel Prize winners and other eminent academic achievements, but the *Live at Leeds* concert by The Who is an equally important part of the university's history.[53]

A mile or two up the Otley Road the vice-chancellorship of Simon Lee at Leeds Metropolitan University (2003-09) witnessed a number of corporate profile-raising initiatives, notably 'partnership' with the Rugby League Challenge Cup and the Headingley Stadium, both of which currently bear the name of the university's

---

52   House of Commons 20th February 2009. Available at www.parliament.the-stationery-office.com/pa/cm200708/cmhansrd/cm080220/halltext/80220h0002.htm [Accessed: 1st June 2009].

53   'Live at Leeds – again' University of Leeds press release 6th June 2006. Available at http://reporter.leeds.ac.uk/press_releases/current/live_at_leeds.htm [Accessed: 15th May 2009].

faculty of sport and education, Carnegie. Through these, and similar, alliances both university and partners would, it was hoped, be enhanced as global brands. How the proprietors of Leeds United must envy Manchester United, arguably their neck-and-neck competitors in the 1960s, in this regard.

Besides this, tourism is a major industry in Yorkshire, generating around £6 billion for the regional economy,[54] principally in the hotel and catering sector. Two tourist attractions in Leeds – the Royal Armouries Museum and the City Art Gallery drew well over a quarter of a million visitors each in 2006[55] but as an aspirational city Leeds must continue to reinvent itself. One reinvention of the city currently mooted is as an international venue for concerts and mega-events. This project bears some of the hallmarks of contemporary Yorkshire politics: it was promoted by the *Yorkshire Evening Post*, who in 2006 conscripted the bands Kasabian and Faithless to speak in support of their 'Leeds Needs an Arena' campaign, launched in September of that year,[56] and it provoked inter-city political strife between Leeds and Sheffield, which already had an arena. Sheffield MPs argued that one of the funding bodies, Yorkshire Forward, should not finance projects which favoured one part of the region over another. The dispute rumbled on through late 2008 into 2009.[57] The arena, which is due to be built at Brunswick Terrace in the city-centre, will, one expects, be managed by out-of-town professionals who will take accommodation in upmarket riverside apartments or houses in the Yorkshire dales. For building work, local workers will very likely compete with workers from Eastern Europe; only the low paid jobs – for cleaners, stewards, ticket-rippers and so on – can be expected to be filled from the local labour market.

Meanwhile the gulf between the polity and the low paid/unemployed occupants of that labour market seems at the time of writing to be wider than ever. A national furore, led by the *Daily Telegraph*, broke out during May/June of 2009 over the accommodation expenses claimed by Members of Parliament and brought MPs from all the major parties into public disgrace. This furore obliterated issues of poverty. Indeed, it was a given and politicians were now reminded of the severity

---

54   Rhodri Thomas (ed.) *Managing Regional Tourism* Ilkley: Great Northern Books 2009 p. 1.

55   ibid p. 5.

56   See Lets get the rock rolling. *Yorkshire Evening Post,* 22nd September 2006. Available   at:http://www.yorkshireeveningpost.co.uk/leeds-needs-an-arena/Lets-get-the-rock-rolling.1783802.jp [Accessed: 12th May 2009] and Andrew Hutchinson. Top bands in plea for Leeds arena. *Yorkshire Evening Post,* 22nd December 2006. Available at: http://www.yorkshireeveningpost.co.uk/leeds-needs-an-arena/Top-bands-in-plea-for.1943770.jp [Accessed: 12th May 2009].

57   See Paul Robinson. Leeds v Sheffield arena row erupts. *Yorkshire Evening Post,* 17th November 2008. Available at: http://www.yorkshireeveningpost.co.uk/news/Leeds-v-Sheffield-arena-row.4699636.jp [Accessed: 14th May 2009] and New war of words over Leeds arena. *Yorkshire Evening Pos,* 2nd January 2009. Available at: http://www.sheffieldtelegraph.co.uk/headlines/New-war-of-words-over.4838070.jp [Accessed: 14th May 2009].

with which benefit fraud among the socially excluded was routinely treated. It is something of a paradox when the comfortable and powerful are trapped like rabbits in the glare of media headlights. Disciplines of state control and welfare surveillance are all of a sudden turned upon the politician themselves. Yvette Cooper, Fabian Hamilton and George Mudie were all cited for excessive claims and/or for dubiously defining their second homes.[58] For many this episode removed any remaining legitimacy from the things they may have said. In May of 2009 the *Daily Telegraph* reported that Mudie had claimed for nearly £17,000 worth of furniture and fittings for his second home in London that had, nevertheless, been delivered to his home in Leeds. In a ruthless contextualisation the paper pointed out that Mr Mudie had 'been at the forefront of parliament's attacks on bank bosses and hedge fund managers in recent months through his role as a member of the all-party treasury select committee'.[59] It seemed that, like a large number of MPs in the Great Expenses Scandal of May-June 2009, Mudie, spokesman for forgotten Leeds and critic of the financial elite, was finished; but Pilger's 'single ideology business state' was not. So, as Paul Chatterton argues:

> we can mark out progressive and regressive elements in the city. As Doreen Massey says, not all places are victims and not all places are worth defending. There are privileged places which hold much power and investment (the financial district, the universities), and there are forgotten places (outer estates) which are stuck in a downward spiral. These are signs of a divided, uncaring city.[60]

Acknowledgements

Thanks to Rhod Thomas, Julie Harpin, Simon Gunn and Paul Marchant for help in the writing of this chapter.

---

58   See *Independent on Sunday* Newsweek supplement: The IOS Guide to our MPs, Their Pay and Those Expense Claims. 24th May 2009, pp. v-vi.

59   Gordon Rayner. George Mudie claimed £62,000 over four years: MPs' expenses. *Daily Telegraph,* 18th May 2009. Available at: http://www.telegraph.co.uk/news/newstopics/mps-expenses/5340103/George-Mudie-claimed-62000-over-four-years-MPs-expenses.html [Accessed: 2nd June 2009].

60   Paul Chatterton. Talking to the Past .... p. 35.

Index